Climate, Society and Elemental Insurance

In this book, world-leading social scientists come together to provide original insights on the capacities and limitations of insurance in a changing world.

Climate change is fundamentally changing the ways we insure, and the ways we think about insurance. This book moves beyond traditional economics and financial understandings of insurance to address the social and geopolitical dimensions of this powerful and pervasive part of contemporary life. Insurance shapes material and social realities, and is shaped by them in turn. The contributing authors of this book show how insurance constitutes and is constituted through the traditional elements of earth, water, air, fire, and the novel element of big data. The applied and theoretical insights presented through this novel elemental approach reveal that insurance is more dynamic, multifaceted, and spatially variegated than commonly imagined.

This book is an authoritative source on the capacities and limitations of insurance. It is a go-to reference for researchers and students in the social sciences – particularly those with an interest in economics and finance, and how these intersect with geography, politics, and society. It is also relevant for those in the disaster, environmental, health, natural, and social sciences who are interested in the role of insurance in addressing risk, resilience, and adaptation.

Kate Booth is a human geographer, specializing in the field of critical insurance studies. She is particularly interested in the economic and social geographies of insurance in a changing climate, and implications for inequality and inequity. Kate has also worked on projects looking at sense of place, and the role of arts and culture in urban regeneration. Her work is published in journals such as *Progress in Human Geography*, *Environment and Planning A: Economy and Space*, *Environment and Planning E: Nature and Space*, *Urban Studies*, and *Qualitative Inquiry*.

Chloe Lucas is a human geographer at the University of Tasmania. A communications specialist, she began her career making documentaries about science and landscape history for the BBC. Chloe's research explores the values and experiences underlying different social responses to climate change, and identifies pathways to more empathetic and inclusive climate conversations. Her recent work focusses on how communication and cultural context drives social adaptation to extreme weather events, and can be found in journals including *Climatic Change, Environment and Planning E: Nature and Space, Geographical Research*, and *WIRES Climate Change*.

Shaun French is an Associate Professor in Economic Geography at the University of Nottingham. He focuses on the geographies of economic practice and knowledge, specifically financial services and money, socially responsible investment, and financial centres. As part of the University's Rights Lab, he is developing new work on debt, vulnerability, and anti-money laundering.

Climate, Society and Elemental Insurance

Capacities and Limitations

Edited by
Kate Booth, Chloe Lucas,
and Shaun French

Routledge
Taylor & Francis Group

LONDON AND NEW YORK

First published 2022
by Routledge
4 Park Square, Milton Park, Abingdon, Oxon OX14 4RN

and by Routledge
605 Third Avenue, New York, NY 10158

*Routledge is an imprint of the Taylor & Francis Group,
an informa business*

British Library Cataloguing-in-Publication Data
A catalogue record for this book is available from the British Library

Library of Congress Cataloging-in-Publication Data
Names: Booth, Kate, editor. | Lucas, Chloe, editor. |
French, Shaun, editor.
Title: Climate, society and elemental insurance : capacities and
limitations / Kate Booth, Chloe Lucas and Shaun French.
Description: Abingdon, Oxon ; New York, NY : Routledge, 2022. |
Includes bibliographical references and index.
Identifiers: LCCN 2021058554 (print) | LCCN 2021058555 (ebook) |
ISBN 9780367743864 (hbk) | ISBN 9780367743871 (pbk) |
ISBN 9781003157571 (ebk)
Subjects: LCSH: Insurance--Social aspects. | Liability for
environmental damages. | Climatic changes. | Sustainability.
Classification: LCC HG8026 .C58 2022 (print) |
LCC HG8026 (ebook) | DDC 368--dc23/eng/20220214
LC record available at https://lccn.loc.gov/2021058554
LC ebook record available at https://lccn.loc.gov/2021058555

ISBN: 978-0-367-74386-4 (hbk)
ISBN: 978-0-367-74387-1 (pbk)
ISBN: 978-1-003-15757-1 (ebk)

DOI: 10.4324/9781003157571

Typeset in Times New Roman
by KnowledgeWorks Global Ltd.

An electronic version of this book is freely available, thanks to the
support of libraries working with Knowledge Unlatched (KU). KU is
a collaborative initiative designed to make high quality books Open
Access for the public good. The Open Access ISBN for this book is
9781003157571. More information about the initiative and links to the
Open Access version can be found at www.knowledgeunlatched.org.

Contents

Tables

Figures

Contributors

Kate Booth, PhD, is a human geographer at the University of Tasmania, specializing in the field of critical insurance studies. She is particularly interested in the economic and social geographies of insurance in a changing climate, and implications for inequality and inequity. Kate has also worked on projects looking at sense of place, and the role of arts and culture in urban regeneration. Her work is published in journals such as *Progress in Human Geography, Environment and Planning A: Economy and Space, Environment and Planning E: Nature and Space, Urban Studies*, and *Qualitative Inquiry.*

Eliza de Vet's, PhD, research interests include individual and household experiences of weather, climate (change), and natural hazard-related disasters, with particular interest in mitigation and adaptation strategies. Based at the University of Wollongong, her key publications relate to everyday weather-related cultural practices and environmentally sustainable thermal comfort strategies. Recently, her research has extended to the role of home and content insurance in household disaster resilience, publishing on building regulations and under-insurance, insurance as an unmitigated disaster, and insurance as essential tool for long-term household recovery.

Rebecca Elliot, PhD, is an Assistant Professor of Sociology at the London School of Economics. Her research examines the intersections of environmental change and economic life, as they appear across public policy, administrative institutions, and everyday practice. She is the author of *Underwater: Loss, Flood Insurance, and the Moral Economy of Climate Change in the United States* published in 2021.

Dr. Christine Eriksen, PhD, University of Wollongong, gained international research recognition by bringing human geography, social justice, and natural hazards into dialogue. With a particular interest in social dimensions of disasters, her widely published work examines social vulnerability and risk adaptation in the context of environmental history, cultural norms, and political agendas. She is the author of two books:

Alliances in the Anthropocene: Fire, Plants and People (2020) and *Gender and Wildfire: Landscapes of Uncertainty* (2014). Dr. Eriksen joined the Center for Security Studies at ETH Zürich, Switzerland in August 2020. Prior to that, she worked for 13 years as an academic at the University of Wollongong, Australia. Her work has focused on case studies in multiple countries and continents, including Australia, North America, Europe, and Africa.

Shaun French, PhD, is an Associate Professor in Economic Geography at the University of Nottingham. He focuses on the geographies of economic practice and knowledge, specifically financial services and money, socially responsible investment, and financial centres. As part of the University's Rights Lab, he is developing new work on debt, vulnerability, and anti-money laundering.

Kevin Grove, PhD, is an Associate Professor of Geography at Florida International University. His research explores the politics of disaster management and resilience in the Caribbean and North American cities. His work has been published in a variety of academic journals, including the *Annals of the American Association of Geographers, Environment and Planning D: Society and Space, Progress in Human Geography*, and *Security Dialogue*, and he is the author most recently of *Resilience* (Routledge) and co-editor (along with David Chandler and Stephanie Wakefield) of *Resilience in the Anthropocene: Governance and Politics at the End of the World* (Routledge).

Olli Hasu is a doctoral student at Tampere University, Finland. His work focuses on climate change, the infrastructure character of new information technologies, and index insurance.

Mark Kammerbauer's, PhD, field of expertise covers risk and resilience in the built environment under consideration of socio-cultural issues, governance, and vulnerability. Over the past 15 years, he has researched cases of disaster recovery and reconstruction in the United States, Germany, Australia, and the Philippines. Mark held research and teaching positions at international higher learning institutions, including the University of Queensland School of Geography, Planning and Environmental Management, Lund University Centre for Sustainability Studies, Tulane University Center for Bioenvironmental Research, and Technische Universität München. Mark is the Director of Nexialist Agency for Research and Communication and writes regularly for urban planning, urban design, and architecture publications.

Ken Klein, PhD, is the Lewis and Hermione Brown Professor of Law at California Western School of Law. A wildfire survivor himself, he teaches and publishes on insurance issues arising in the wake of natural disasters. He is a consumer representative to the National Association of Insurance

Commissioners (NAIC) and has presented on issues of insurance and wildfire to the NAIC, the Federal Home Finance Administration, and to HopeNow, an organization of mortgage lenders/servicers. For his thousands of hours of volunteer work on-site in the wake of natural disasters, he has been awarded the State Bar of California's highest award for pro bono legal services.

Turo-Kimmo Lehtonen, PhD, is a Professor of Sociology at Tampere University, Finland. His present work centres on two different areas; one of which is insurance and the management of uncertainty, and the other the role of waste in contemporary life. In addition, Lehtonen has written extensively on social theory. His recent work includes papers in the journals *Political Theory*, *Cultural Studies*, *Distinktion*, *Theory, Culture & Society*, *Valuation Studies*, and *Economy and Society*.

Chloe Lucas, PhD, is a human geographer at the University of Tasmania. A communications specialist, she began her career making documentaries about science and landscape history for the BBC. Chloe's research explores the values and experiences underlying different social responses to climate change, and identifies pathways to more empathetic and inclusive climate conversations. Her recent work focusses on how communication and cultural context drives social adaptation to extreme weather events, and can be found in journals including *Climatic Change*, *Environment and Planning E: Nature and Space*, *Geographical Research*, and *WIRES Climate Change*.

Liz McFall, PhD, is the Director of Data Civics and Chancellor's Fellow based in the Edinburgh Futures Institute and Sociology at the University of Edinburgh. She co-edited *Markets and the Arts of Attachment* (2017), wrote *Devising Consumption: Cultural Economies of Insurance, Credit and Spending* (2014) and *Advertising: A Cultural Economy* (2004) and edits the *Journal of Cultural Economy*. She is co-founder of AWED, a collective that make films and installations exploring the orchestration of civic sentiments and data techniques, most recently *Closes and Opens: A History of Edinburgh's Futures* – available on YouTube.

Scott McKinnon, PhD, is an oral historian and geographer at the Australian Centre for Culture, Environment, Society and Space (ACCESS), University of Wollongong. His research explores the social dimensions of disaster, including processes of disaster commemoration, and the impacts of disaster on marginalized social groups. He is co-editor (with Margaret Cook) of *Disasters in Australia and New Zealand: Historical Approaches to Understanding Catastrophe* (published 2020). Scott has also published extensively on histories and geographies of sexuality, including *Gay Men at the Movies: Cinema, Memory and the History of a Gay Male Community* (2016).

Pat O'Malley, PhD, is a Distinguished Honorary Professor in the School of Sociology at the Australian National University and an Adjunct Research Professor in Sociology at Carleton University in Canada. Previously, he was Professorial Research Fellow in Law at the University of Sydney. Most research over the past 25 years has focused on risk as a form of governance in criminal and civil justice, disaster management, and insurance. Recent research into insurance has focused on the role of the insurance industry in promoting and enacting risk-based fire prevention regimes in urban settings.

Nick Osbaldiston, PhD, is a Senior Lecturer in Sociology at James Cook University, Australia. His research focusses on three areas: internal migration, adaptation to environmental risks, and coastal research. He has recently published the monograph *Towards a Sociology of the Coast* (published 2018) and prior to that *Seeking Authenticity in Place, Culture, and the Self* (published 2012).

Lauren Rickards, PhD, is a Professor in the Centre for Urban Research at RMIT University, Melbourne, where she leads the Urban Futures Enabling Capability Platform and the Climate Change Transformations research program. A human geographer, Lauren's work examines the ideas and discourses that shape how we interact with the environment, notably with issues of climate, land, and food. Lauren is a lead author with the forthcoming IPCC report on Impacts, Adaptation and Vulnerability.

Minna Ruckenstein, PhD, is an Associate Professor at the Centre for Consumer Society Research at University of Helsinki, Finland. She leads an interdisciplinary research group, and is currently directing two large research projects, one focusing on algorithmic culture and the other on rehumanising automated decision-making.

Maiju Tanninen, PhD, is a doctoral researcher at University of Helsinki and Tampere University. In her ongoing PhD research, she studies the (new) datafication of life insurance markets, focusing on both industry practices and consumers' everyday experiences with the emerging technologies.

Zac Taylor, PhD, is a Research Fellow in the Faculty of Architecture and the Built Environment at Delft University of Technology. They research urban climate risk governance within the financial system and across three contexts (South Florida, Singapore, and the Netherlands). Zac currently a co-convenes the Urban Climate Finance Network, and their recent work can be found in *Environment and Planning A: Economy and Space* and the *Cambridge Journal of Regions, Economy and Society*.

Jonathon Turnbull is a cultural and environmental geographer based at the University of Cambridge. His current research explores the human–animal relations and weird ecologies of the Chernobyl Exclusion Zone with particular reference to dogs and wolves. He also writes on digital ecologies, more-than-human filmic geographies, and bovine geographies.

Christine Wamsler, PhD, is an internationally-renowned expert in sustainable development with focus on disaster risk reduction, climate change adaptation, urban resilience and transformation. She is Professor at Lund University Centre for Sustainability Studies (LUCSUS) in Sweden and Honorary Fellow at the Global Urban Research Centre and the Institute for Development Policy and Management (IDPM) of the University of Manchester, UK. Previously, she worked as co-director of the Societal Resilience Centre. In parallel to her academic research, Christine works on an ongoing basis as a consultant for different governmental and non-governmental organisations.

Travis Young, PhD, is a postdoctoral associate in the Department of Environment and Sustainability at the University at Buffalo. Before completing his doctorate at Penn State University, he worked as a disaster recovery planner in the Texas Gulf Coast following Hurricane Ike. Through ARC Discovery and National Science Foundation grants, Travis worked with Kate Booth and Chloe Lucas exploring the connections between bushfire risk and home insurance in Australia.

Acknowledgement

This book was supported partially by the Australian Government through the Australian Research Council's Discovery Projects funding scheme (project DP170100096). The views expressed herein are those of the authors and are not necessarily those of the Australian Government or Australian Research Council.

1 Introduction

Kate Booth

Our climate is being privatised, and with it, our democratic capacity to act in the face of unprecedented risks is being curtailed. Insurance and insurers are central to these changes. The insurance sector, for example, is a major player in global financial markets and is continuously expanding its reach into new places and through novel products. This sector is emerging as the leader in climate risk – the assessor of climate risk, the holder of risk data, and the actor in face of government inaction (Lehtonen 2017). These shifts in governance are refracted in neoliberal reforms pertaining to the withdrawal of the state, the individualisation of risk, and self-responsibilisation (Harvey 2007). This is hardening existing structures of inequality and contributing to new patterns of inequity (e.g. Booth & Kendal 2020). The gravitas of this nexus of power, crisis, and injustice is contributing to a growing body of critical insurance research within the social sciences.

Climate change is fundamentally changing the ways we insure, and the ways we think about insurance, and there is an urgent need to build knowledge on the capacities and limitations of insurance in this regard. Thus, our aim in this book is, for the first time, to bring together original, world-leading social research on insurance. In garnering the thoughts and insights of contributing human geographers, sociologists, built environment experts, and others, our impetus is to contribute to growing this body of research and charting the terrain for future inquiry.

The origins of insurance in actuarialism and finance influence how it is approached in research and practice. Most texts on insurance are located within the fields of economics and finance, focusing on key concepts and principles, legal frameworks, and explanations of the insurance industry. Quantitative methods dominate with an assumption of individual rational decision-making, and this has shaped insurance as a scientific object – a tool to be applied and a thing to be measured in its effectiveness (or lack thereof). In the social sciences, however, such perspectives have been critiqued, unsettled, and complicated. As well as being actuarial and financial, insurance is affective, emotional and normative, and constituted within everyday life as well as the more abstract spheres of global finance (Pike & Pollard 2010).

DOI: 10.4324/9781003157571-1

As we show in this book, insurance is also *elemental* – it is co-constituted within the dynamic entwining of earth, water, air, fire, and, what we propose as a fifth element, 'big' data.

The shifting ontological terrain that is insurance is reflected in diversification of epistemological approaches in insurance research. The methods employed include the qualitative and discursive, as well as quantitative, and as we see this diversity as representing the 'bricolage qualities' of insurance (French & Kneale 2015, p. 17), an objective of this book is inclusivity. Insurance takes on different forms through different epistemological practices, and its constitution is spatially and temporally variegated. Thus, the 14 standalone chapters in this book provide representations of different *insurances* embodying dynamic and unbounded places, technologies, and networks.

To find cohesion in this multiplicity, through the book we follow two narrative threads. One is ontological – we trace how insurance realities constitute and are constituted through the elements of earth, water, air, fire, and 'big' data. In the creation and evolution of insurance technologies, these elements are configured into forms that fit hegemonic insurance logics. Each is rendered calculable and potentially predictable, stripped of amorphous, mysterious, and spiritual qualities frequently associated with each in other times and other places. Earth becomes soil moisture, water is delineated by volume and flow, air is reduced to wind speed and temperature, fire is defined by frequency and severity. 'Big' data congeals as information and fact after being scrutinised and cleaned to remove mishaps and inconsistencies emergent in everyday life and through technological glitches and limitations. In turn, insurance itself is constituted through these elements because despite the power and prowess of insurance logics, these dynamic and unbounded elements exert force in the design and implementation of insurance technologies. These, at times, wild elements constitute ideas of insurability and non-insurability with excess – excessive wind, rain, temperature – in part determining where the capacities and limitations of insurance may lie. Or, more obviously, the need for certain types of insurance at all.

The other narrative thread is epistemological. In relation to insurance, we loosely map these elements onto different spatialities – earth and Kantian space (Ingold 2008), air and forgotten space (Irigaray 1999), water and fluid space, fire and fire space (Law & Mol 2001), and 'big' data and spatial imaginaries (Lobo-Guerrero 2014; Watkins 2015). Mobilising or recognising different spatialities can provide new insights, specifically the multiple, sometimes conflicting, realities of insurance (Booth 2021). For example, the idea of fluid space can lead to articulations pertaining to the internal dynamics of insurance. Insurance technologies evolve over time (Ewald 1991), they are variously constituted in different places (Booth & Kendal 2020), *and* the same type of insurance – even a single insurance policy – is variously constituted through both time and space through emotions, morality, and place. As Booth (2021, p. 1306) observes, such dynamism,

...points towards modes of knowing that are open and sensitive to movement and mutability within insurance, and what this means for insurance praxis – the enactment or realisation of these theoretical insights through practice and politics.

By making these ontological and epistemological connections between the sections and chapters of this book, we reveal key themes in contemporary research that signal the capacities and limitations of insurance. This includes how dualisms associated with insurance – rational/emotional, technical/political, individual/collective – are unsettled and disrupted, providing new avenues for understanding and research. We synthesise and elucidate these insights in the book's concluding chapter.

Between now and then, however, we begin our elemental exploration of insurance in Section 1 with 'Earth'. Lauren Rickards begins by delving into insurance and geo-engineering 'fixes' to climate change, such as soil carbon sequestration. She observes that insurance and geo-engineering both posit culturally embedded narratives of *'If this continues, then...'* and *'What if...?'* Insurance provides an imagining (Ewald 1991) that addresses disasters and risks associated with climate change with the hopeful, *'What if I had insurance?'* Yet, despite limits to insurability, insurance remains unambitious when it comes to claiming or pursuing a transformed future. This imagining of business-as-usual is interlinked with the possibility of a climate stabilised through geo-engineering, and acts to 'dampen' other narratives that forewarn of the implications of unmitigated climate change. Rickards concludes with a call to consider a different kind of *'If only...'* – if only we could escape Modernist narratives and instead look to how we could relate with reciprocity and solidarity to a changing Earth.

In the next two 'Earth' chapters, Olli Hasu, Turo-Kimmo Lehtonen, and Kevin Grove turn their attention to forms of parametric or index insurance in the developing world. Hasu and Lehtonen critically explore the *Global Index Insurance Facility* (GIIF) project – a project led by the World Bank and aimed at providing financial services for smallholder farmers. Adding to Rickards' observations about insurance logics, they describe the process by which soil is turned into an index, how this index is then applied, and ultimately how this integrates soil into global financial markets. What emerges is a technology that claims innovative significance yet appears to be a traditional form of insurance premised in large part on archival-statistical knowledge. The index remains dependent on fixed and unchanging environmental indicators and does not or cannot accommodate dynamic variables. Despite climate change being at the heart of the GIIF project, this form of insurance appears unresponsive to this phenomenon.

From the plantation state of Dominica in the Caribbean, Kevin Grove also tackles index insurance within the context of ex-ante risk management (EARM). He observes how new technologies like catastrophe models and parametric insurance appear to disrupt the idea that there are 'earthly' limits to the capacities of insurance. Again, reflecting the power of insurance logics, EARM pushes governments to 'think like an insurer.' The political ramifications of this normalisation illustrate the place-based constitution of insurance and, thus, its capacities and limitations. Within the context of promises of sovereignty, autonomy and independence, and an end to the structural violence of plantation economies, the moral imperative of EARM for financial self-sufficiency in an increasingly volatile climate, reinforces rather than dissolves social and racial inequalities and inequities.

To begin Section 2, 'Water', Rebecca Elliot reflects on flood insurance in the United States and the United Kingdom, and the moves to minimise or eliminate cross-subsidisation that has enabled those in high-risk areas to afford to live and build there (with insurance) through the redistribution of money from those paying insurance in low-risk areas. In the United States, reducing cross-subsidisation is argued to achieve more equitable outcomes given improved capacity to ascertain and rate flood risk. In the United Kingdom, similar reductions are motivated by a redistribution of prudence and responsibility that comes with clearer risk 'price-signals.' Elliot identifies a flaw in the logics underpinning this curtailing of cross-subsidisation in both countries – namely that social interdependence always, already necessitates sharing of risk and responsibility irrespective of neoliberal, market-based logics. Thus, the focus could or should be on the nature and desired outcomes of cross-subsidisation rather than blinkered attempts at elimination.

Elliot's observations resonate with the preceding chapters in this section, both of which draw in interviews with households in flood-prone or flood-impacts areas. Chloe Lucas and Travis Young compare the experiences and perceptions of insurance for residents in Hobart, Australia and Houston, United States. Both places are covered by different insurance structures for flooding – private insurance in Australia and the National Insurance Flood Program (NIFP) in the United States, and have different socio-demographic profiles – middle class in Hobart and racially and socially diverse in Houston. Yet, between continents, households share understandings of resilience and the capacities of insurance that are constructed through uncertainty and dependent on varying levels of political empowerment. These understandings are in stark contrast to the resilience and insurance discourses of government and the insurance sector, which position insurance as essentially unproblematic and apolitical. Mark Kammerbauer and Christine Wamsler also point to disconnects between government discourse and householder experiences of the 2013 flooding in Deggendorf, Germany. Many homes were inundated not only by water, but also by heating oil from ruptured piping. In this context, uninsured households recovered more quickly with the support of government assistance than those with insurance, despite European Union aspirations for self-responsibilisation. A lack of integration in flood management suggests that similar ad hoc outcomes will continue, some of which will be maladaptive (O'Hare et al. 2015).

Section 3, 'Fire', begins with another study that draws on interviews with residents in at-risk places and again brings attention to the fallibility of insurance logics – in this case, in the fire-prone Blue Mountains and Kangaroo Valley of New South Wales, Australia. Scott McKinnon, Christine Eriksen, and Eliza de Vet unsettle hegemonic discourses pertaining to the presence and absence of insurance – frequently represented as a normative dichotomy between having insurance (good) and not having insurance (bad). In examining the idea of value, McKinnon and colleagues observe that in the eyes of bushfire survivors, adequate insurance coverage is defined in more than monetary terms, and that time and emotional and physical labour are key currencies to recovery beyond the financial possibilities of insurance. The notion of adequate insurance is thus dependent on socio-cultural context.

Kenneth Klein also delves into the adequacy and availability of insurance in fire-prone landscapes in Australia and the United States, posing the question, 'as climate change continues, is fire insurable?' He argues that even now when insurance for wildfire remains broadly accessible and most households want to be insured and many assume they are adequately covered, despite best intentions, most are not. Klein observes that there is no publicly available data on what and how insurers know about adequate levels of cover in the face of fire and, problematically, the onus for underinsurance lies with the householder. Opening with the ongoing ramifications of the Grenfell Tower fire in London, Pat O'Malley similarly critiques the institutional configurations that constitute flammability, technical error, and insurance. Drawing on a case study of fire/building politics in Australia that followed the Grenfell incident, he reflects upon the international conglomerates, including construction and insurance industries, that inform and influence the debate. Specifically, O'Malley presents a critique of the concept of sustainability that is frequently mobilised unproblematically in decision-making, yet carries a level of complexity that undermines coherent responses with subsequent implications for insurance and insurability.

In the penultimate Section 4, 'Air' is the element in focus. Zac Taylor analyses insurance-linked securitisation as constituted through hurricane wind risk in Florida, United States and risk-brokers in Singapore. As a pivotal part of this evolving relational network, re/insurance faces 'headwinds' of change – the risk of sea-level rise to Florida's coastal communities, the economic sustainability of Singapore's financial sector, and the tenability of long-term insurability. Yet, this sector remains a powerful force in defining and managing climate uncertainty – at least for now. Networks of global finance often remain abstruse for those living with risks on a day-to-day basis, and in the next chapter, Nick Osbaldiston provides an account of how households in cyclone-prone northern Australia make sense of insurance. Residents express a range of emotions that complicate and at times contradict and obscure hegemonic financial discourses and the framing of customers as rational and calculating. For some, a desire to insure acts in concert with feelings of mistrust in insurers and leads to decisions to insure that are laced with uncertainty. For others, especially renters, previous experience is disasters and other disruptive events that lead to a sense of optimism and confidence for living without insurance.

Christine Eriksen and Jonathon Turnbull conclude the section looking at Chernobyl, the site of the 1986 nuclear powerplant explosion. Recent wildfires have activated radioisotopes held within the soil and vegetation, presenting the possibility of contaminated plumes billowing across Europe. This airborne risk transcends national borders and appears to supersede liability in the obscurity of its cause. In drawing on concepts of affective atmospheres and volumetric sovereignty, Eriksen and Turnbull signpost implications of insurance and disaster scholarship.

Our last section concerns the contemporary element of 'Big Data,' and the two final chapters contained herein contribute to a growing body of work that is following the rollout of, and experimentation with big data and analytics in the insurance sector. Drawing also upon interviews with customers and potential customers, Maiju Tanninen, Turo-Kimmo Lehtonen, and Minna Ruckenstein explore attitudes and experiences towards new insurance technologies that embody sensor-generated and self-tracking data. They describe how customers negotiate a 'leap of faith' in embracing this new kind of insurer-insuree relationship based on the sharing of intimate, personalised data. Like Osbaldiston's observations, emotions imbue this relationship, with customers expressing interest and curiosity, while also experiencing doubt, hesitation, and uncertainty.

Liz McFall concludes the chapter sections, by bringing 'big' data back to earth. Tracking the architecture of insurance company buildings in the United Kingdom and the United States, in tandem with developments in technologies, McFall illustrates how Insurtech and 'big' data are contributing to urban form. In the early days of insurance, companies worked to build trust and a sense of solidity through an assemblage of branding devices including monumental stone buildings in Central Business Districts (CBDs). Emphasising creativity and innovation, companies experimenting with Insurtech now inhabit retrofitted, mobile spaces in gentrifying areas. These buildings signify disruption and change, rather than the dependability and certainty of older insurance offices. Despite the contradictions, schisms and tensions evident in the capacities of insurance illustrated in proceeding chapters, McFall reminds us that the seemingly distant and abstract power and prowess of insurance and insurers remain a concrete presence within our built environments.

In this summary of the 14 chapters, we have not unpacked a significant feature of this body of work – the entwinning of elements in the co-constitution of insurance. Many authors speak to multiple elements even if we have categorised them in relation to one. For example, Eriksen and Turnbull describe the air-borne distribution of radiation from Chernobyl as it is released from the earth and vegetation through wildfire. Here, air, earth, and fire combine to produce an uncontainable and uninsurable event well beyond the original 1986 disaster. In her discursive account of the relationship between Insurtech and its companies' buildings, McFall also provides a graphic account of the enmeshment of earth, air, and data. Airy contemporary insurance offices stand in stark contrast to the weighty stone buildings of the 19th-century insurance companies. This entwinning appears to warrant future attention in understanding the elemental co-constitute of insurance in a climatically shifting world.

References

Booth, K 2021, 'Critical insurance studies: Some geographic directions', *Progress in Human Geography*, vol. 45, no. 5, pp. 1295–310.

Booth, K & Kendal, D 2020, 'Underinsurance as adaptation: Household agency in places of marketization and financialization', *Environment and Planning A: Economy and Space*, vol. 52, no. 4, pp. 728–746.

Ewald, F 1991 'Insurance and risk', In, Burchell, G, Gordon, C. & Miller, P. (eds) *The Foucault Effect: Studies in Governmentality*. Hertfordshire, UK: Harvester Wheatsheaf, pp. 197–210.

French, S & Kneale, J 2015, 'Insuring biofinance: Alcohol, risk and the limits of life', *Economic Sociology*, vol. 17, no. 1, pp. 16–24.

Harvey, D 2007, *A Brief History of Neoliberalism*, Oxford, UK: Oxford University Press.

Ingold, T 2008, 'Bindings against boundaries: entanglements of life in an open world', *Environment and Planning A*, vol. 40, pp. 1796–1810.

Irigaray, I 1999, *The Forgetting of Air in Martin Heidegger*, London, UK: Athlon Press.

Law, J & Mol, A 2001, 'Situating technoscience: An inquiry into spatialities', *Environment and Planning D: Society and Space*, vol. 19, pp. 609–621.

Lehtonen, T 2017, 'Objectifying climate change: Weather-related catastrophes as risks and opportunities for reinsurance', *Political Theory*, vol. 45, no. 1, pp. 32–51.

Lobo-Guerrero, L 2014, 'The capitalisation of 'excess life' through life insurance', *Global Society*, vol. 28, no. 3, pp. 300–316.

O'Hare, P, White, I & Connelly, A 2015, 'Insurance as maladaptation: Resilience and the 'business as usual' paradox', *Environment and Planning C: Government and Policy*, vol. 34, no. 16, pp. 1175–1193.

Pike, A & Pollard, J 2010, 'Economic geographies of financialization', *Economic Geography*, vol. 86, no. 1, pp. 29–51.

Watkins, J 2015, 'Spatial imaginaries research in geography: Synergies, tensions, and new directions', *Geography Compass*, vol. 9, no. 9, pp. 508–522.

Section I

Earth

Earth is one of the classical elements – along with air, water, and fire – and was once understood as foundational to the material world. For the Ancient Greeks, earth spoke of matter and the terrestrial realm, and in other parts of the world in different traditions was symbolized by, for example, the bull and bird (Christian), lotus (Hindus), and house (Aztecs).

Modern science has provided an alternative configuration of the material world – favouring chemical elements comprised of atoms and comprising compounds, and negating place-based specificity. As gases, liquids, and solids, these chemicals are understood as the physical basis of a material world with universalized and universalizing properties.

Such universalism depends, as Law and Mol (2001, p. 609) observe, on 'never asking *where*-questions.' Thinking *where*, brings things – quite literally – back down to *earth*. Back on earth, the material world emerges as localized, manifesting and a manifestation of places with particular soils, bedrock, plant and animal life, culture, structures, politics – the list goes on.

Early research on insurance was limited in its understanding of locality and place, emphasising its constitution through the economic, juridical, political, and moral in the western world (e.g. Ewald, 1991). This is reflected in understandings of economic and financial processes more generally: that the scale and speed of such globalized and globalizing processes can create a sense of disconnection from 'geographic entanglements in space and place' (Pike & Pollard 2010, p. 36).

More recently, place-based insurance research has included studies of insurance-related phenomenon at various scales. For example, in developing regions (e.g. Grove 2012; Johnson 2013; Müller et al. 2017), nation states (e.g. Booth & Tranter 2018; Langley 2006; Lehtonen 2017; Lobo-Guerrero 2011; O'Malley & Roberts 2013), and specific risk-affected areas (e.g. Booth & Harwood 2016; Booth & Kendal 2020; Nance 2015). There is also research that engages with the social dimensions of place in considering patterns of socio-demographic change constituted through insurance and insurability (e.g. Booth & Kendal 2020; Johnson 2015). Of a more earthy nature, there are investigations of agricultural insurance, or rather micro-insurance schemes in which soil moisture and crop failure (amongst other factors) are

DOI: 10.4324/9781003157571-2

used to trigger payouts to subsistence farmers in the developing world (e.g. Johnson 2013; Müller et al. 2017). These types of insurance are being instigated through partnerships between insurers and non-government development organisations or the United Nations. They are frequently tied to prescriptions for seed source, crop type, and agro-chemical use, and thus can be formative in the changing nature of places by facilitating integration within large-scale agricultural systems. On the other hand, micro-insurance purchase in particular places can be informed by local knowledge and traditions, in which the logics of insurance are changed because of, and in concert with place specificities (Johnson 2013).

Thinking does not stop with the idea of coming 'down to earth' in insurance research, as ideas of place raise a different type of question – not only what location, but what kind of *place* or 'what kind of space' (Law & Mol 2001, p. 610). Ingold (2008, p. 1798), for example, observes how Kantian notions of place and space that inform modern scientific thinking locate people on the outside of earth: 'They do not... live *within* the world but *upon* its outer surface.' People are represented as in some sense removed from what goes on around them, with places understood as backdrops or stages for human activity. It is possible, within this framing, to map particular phenomenon as place-based according to longitude and latitude.

In subsequent sections, we reflect on other conceptualizations of place and space, and how these intersect and inform insurance research and understandings of insurance. In this section, the selected chapters pivot on earthliness constituted through soil, relationality, and locality. In the first chapter, Lauren Rickards emphasises the power of cultural narratives that constitute the insurance imaginary in concert with geo-engineering responses to climate change, such as soil carbon sequestration. Drawing upon Latour (2018), she posits 'the Terrestrial' as a relational challenge to Modernist logics of insurance discourse and practice. Olli Hasu and Kimmo-Turo Lehtonen also cast a critical eye over the stories told in relation to insurance in their chapter – in this case how soil is accounted for through index insurance targeted at farmers in the developing world and justified on the basis of climate risks. They conclude, that despite claims of innovation and significance, this insurance type remains quite traditional in form and thus, unresponsive to dynamic climate variables. A similar, 'schismed' insurance imaginary (Ewald 1991) inhabits the rollout of index insurance in the Caribbean. In the final chapter, Kevin Grove illustrates how former plantation states are encouraged to 'think like an insurer' to deliver financial sustainability in the face of disasters, yet through this logic experience a reinforcement and intensification the racialized governance that undercut post-colonialist gains.

Insurance logics attempt to hold different parts of the world in place as 'backdrops' before which insurance performs it's abstract and universalized capabilities. However, exploring the expansion of global capital through insurance in different locations reveals the stark, and at times brutal, co-constitution of place and insurance.

References

Booth, K & Harwood, A 2016, 'Insurance as catastrophe: A geography of house and contents insurance in a bushfire prone area', *Geoforum*, vol. 69, p. 44–52.

Booth, K & Kendal, D 2020, 'Underinsurance as adaptation: Household agency in places of marketization and financialization', *Environment and Planning A: Economy and Space*, vol. 52, no. 4, pp. 728–746.

Booth, K & Tranter, B 2018, 'When disaster strikes: Under-insurance in Australian households', *Urban Studies*, vol. 55, no. 14, pp. 3135–3150.

Ewald, F 1991, 'Insurance and risk', In, Burchell, G, Gordon, C & Miller, P (eds.), *The Foucault effect: Studies in governmentality*, Hertfordshire, UK: Harvester Wheatsheaf, pp. 197–210.

Grove, K 2012, 'Preempting the next disaster: Catastrophe insurance and the financialization of disaster management, *Security Dialogue*, vol. 43, no. 2, pp. 139–155.

Ingold, T 2008, 'Bindings against boundaries: Entanglements of life in an open world, *Environment and Planning A*, vol. 40, pp. 1796–1810.

Johnson, L 2013, 'Index insurance and the articulation of risk-bearing subjects', *Environment and Planning A: Economy and Space*, vol. 45, pp. 2663–2681.

Johnson, L 2015, 'Catastrophic fixes: Cyclical devaluation and accumulation through climate change impacts', *Environment and Planning A: Economy and Society*, vol. 47, pp. 2503–2521.

Langley, P 2006, 'The making of investor subjects in Anglo-American pensions', *Environment and Planning D: Society and Space*, vol. 24, pp. 919–934.

Latour, B 2018, *Down to earth: Politics in the new climatic regime*, London: John Wiley & Sons.

Law, J & Mol, A 2001, 'Situating technoscience: An inquiry into spatialities', *Environment and Planning D: Society and Space*, vol. 19, pp. 609–621.

Lehtonen, T 2017, 'Domesticating insurance, financializing family lives: The case of private health insurance for children in Finland', *Cultural Studies*, vol. 31, no. 5, pp. 685–711.

Lobo-Guerrero, L 2011, *Insuring security: Biopolitics, security and risk*, Oxon, UK: Routledge.

Müller, B, Johnson, L & Kreuer, D 2017, 'Maladaptive outcomes of climate insurance in agriculture', *Global Environmental Change*, vol. 46, pp. 23–33.

Nance, E 2015, 'Exploring the impacts of flood insurance reform on vulnerable communities', *International Journal of Disaster Risk Reduction*, vol. 13, pp. 20–36.

O'Malley, P & Roberts, A 2013, 'Governmental conditions for the economization of uncertainty: Fire insurance, regulation and insurance actuarialism', *Journal of Cultural Economy*, vol. 7, no. 3, pp. 253–272.

Pike, A & Pollard, J 2010, 'Economic geographies of financialization', *Economic Geography*, vol. 86, no. 1, pp. 29–51.

2 Insurance and geoengineering

From the delusional to the terrestrial?

Lauren Rickards

Insurance as utopian narrative

Insurance is inherently future-oriented. Acknowledging the need for insurance is about acknowledging the fact that things may go wrong in the future, that disasters may occur, that our current life may be disrupted. At the same time, in response to this dystopian vision, insurance seeks to reassure; it offers a Plan B, a compensatory measure to speed our recovery to an imagined disruption. In this way, the insurance industry flits between two of the three classic narratives about the future, a triad of utopian stories[1] characterised, respectively, by the phrases *'If this continues, then.'*, *'What if?'*, and *'If only.'* Insurance warns us that *if* the volatility and riskiness of the world continues – as it will increasingly do under climate change – *then* all of us will likely face periods of disruption, loss, and damage. Once we are forewarned, it posits a solution: *What if* you had insurance? *What if* you had a Plan B? A promise is dangled: our future would be difficult but not destitute, different but not disastrous.

This triad of narratives about the future – *'If this continues, then.'*, *'What if?'*, and *'If only.'* – are not only stylistic devices used by writers of utopian science fiction, but are a reflection of a triptych that Bridgit Schneider (2017) traces to eschatological depictions of the future within Christianity. Capturing messages of the prophets, they represent warnings, speculations, and promissory visions. Schneider argues that a similar triptych features in IPCC depictions of possible futures under their Representative Concentration Pathways: a dark climate future with out-of-control atmospheric greenhouse gas concentrations, a moderate climate future, and a least disastrous climate future secured by rapid and immediate mitigation and, notably, widespread use of 'negative emissions technologies' (NETs) – that is, carbon removal geoengineering techniques. Jumping genres, Schneider also suggests that the same sort of triad features in a 1979/1982 cartoon *A Short History of America* by Robert Crumb. The cartoon depicts the gradual replacement of 'natural landscapes' with urban sprawl, and three potential futures: the Ecological Disaster that arguably awaits *if* urbanisation continues unchecked; the ecomodernist Fun Future that capitalists suggest we can

DOI: 10.4324/9781003157571-3

grasp *if* we are agile and creative enough; and the Ecotopian Solution that is possible *if only* we change direction and return to grounded, simple living.

Latour (2018, p. 52) offers another triad of utopias with his description of the Local, the Global, and the Out-of-This-World attractors that he suggests orient modernisation. The classic modernisation front is from the Local (the traditional) to the Global (the advanced), while the Out-of-This-World option has emerged as a form of radical climate change denial by those who fantasise that they are 'explicitly outside of all worldly constraints, literally *offshore*, like a tax haven' (p. 36). There is a resemblance here with the triad of utopian narratives above. The Local's focus on securing a refuge resonates with the downbeat '*If this continues, then...*'; the Global's focus on endless solutions and upwards progress resonates with the more upbeat '*What if...?*'; and the Out-of-This-World's focus on escape resonates with the longing of '*If only...*' (even as it differs radically from Crumb's Ecotopian Solution). Latour argues that none of these utopias are sufficient for our current times. Thus, the fact that they arguably structure much current thinking about insurance – and, as we will discuss, geoengineering – is a concern. At the same time, elements of geoengineering – notably soil carbon sequestration – offer glimpses of what Latour posits as a way out: a turn to a different attractor, 'the Terrestrial,' which sidesteps modernisation's blind pursuit of human progress by returning to e/Earth, grounding us in the reality of the need to respect and work with, not exploit or retreat from, the planet we are part of.

In this chapter, I explore the relationships between insurance and geoengineering to better understand their shared logics, entanglements, and stance on the future. As much as insurance is a formal, commercial product and political technology (one increasingly promoted as a climate change tool), I am interested in insurance as a mindset – what Ewald (1991) calls an 'insurantial imaginary.' As an imaginary, insurance absorbs and reflects narratives of the sort above, often in problematic ways. Yet, like geoengineering, it contains seeds of progressive potential. In what follows, I examine the geoengineering-insurance relationship in general and then in terms of terrestrial carbon sequestration, the most earthly of which holds the promise of helping us rewrite our narratives about the future.

Geoengineering as insurance and insurance market

Greenhouse gas mitigation strategies across the world are yet to successfully reduce global heating, throwing the Paris Climate Agreement's call to try to contain average global heating to 1.5°C into question. With modelled pathways to 1.5°C signalling a massive role for NETs (Masson-Delmotte et al. 2018), the growing urgency has triggered a turn to geoengineering as a kind of insurance strategy, as a World Bank blog – *Geoengineering: a possible climate change insurance policy* (McCracken 2009) – illustrates. Since then, the urgency of abating climate change has only increased. So too has interest in two main classes of geoengineering as a kind of insurance strategy: Solar Radiation Management, which acts directly on the Earth's radiation balance to try to alleviate global heating; and NETs/Carbon Removal/ Terrestrial Carbon Management, which seeks to draw carbon down out of the atmosphere and to store it in the terrestrial sphere ('sequestration') to reduce climatic change. Such strategies offer what advocates call a Plan B to the increasingly fallible Plan A of reducing emissions quickly (Fragniere & Gardiner 2016, p. 15). The narrative logic is that *if* emissions continue to rise, *then* we will need geoengineering to slow the pace of global warming to help us cope. Scientists who are pessimistic about mitigation and foresee severe climate change impacts in their home country tend to be most supportive of geoengineering (Dannenberg & Zitzelsberger 2019).

Alongside scientific experimentation and modelling of various geoengineering options – from mirrors in space to Carbon Capture and Storage (geosequestration) – vehement criticisms of geoengineering as a planetary insurance policy have emerged (Anshelm & Hansson 2016). Three stand out. First is that at a fundamental level the very idea of geoengineering is inherently fatalistic (Anshelm & Hansson 2014). Just as insurance has long been resisted as revelling in doom (Ewald 2019), geoengineering's defeatism about greenhouse gas mitigation is angrily rejected by critics as 'treating the symptom over the cause' (Kiehl 2006) and condoning not fighting the deep and unnecessary injustices and harms of climate change. Indeed, some argue that – as with other abuses of emergency rhetoric in response to climate change – geoengineering does not just shrug off but actively worsens climate injustice by opening up a new frontier for profit and geopolitical gain (Markusson et al. 2014).

Those defending geoengineering and insurance in general counter that they are simply being realistic; that it is those denying the inevitability of future disruptions and disasters that are deluded. Drawing on Sigmund Freud, Ewald (2019, p. 138) suggests that 'insurance is merciless' towards 'negation,' by which he means 'a certain manner of relating to reality that consists of denying it at the same moment of its recognition' in order 'to diminish its importance, both individually and collectively.' The insurantial imaginary, in other words, is open to the truth of future risk and promotes proactive and rationally protection. It performs a version of human civilisation in which 'recognizing risk [is] a moral imperative, a principle of rational conduct [...] a civilization that refuses the comfort of ignorance, even shared' (Ewald 2019, p. 140). Such a civilisation is quintessentially Modern, manfully striding towards what Latour (2018) describes as the unreachable Global horizon of complete knowledge, leaving behind those who are wilfully irrational.

Another realm of climate change response shaped by the same rationalist, insurantial imaginary, though only in its dominant risk management guise, is climate change adaptation. Just as insurance pushes against negation, adaptation pushes against 'climate impact denial' (the denial – not of the existence or causes of human-induced climate change – but of the stark risks it poses). More specifically, climate risk management calls upon decision-makers to reduce risk by taking the sensible precaution of investing in commercial insurance. By attaching a price signal to an entity's risk profile, property insurance is presented as *'the* disaster management tool of choice' (Booth & Tranter 2018, p. 2) and a key enabler of longer-term climate change adaptation. Driving acceptance of insurance as a form of adaptation is raising awareness of the growing 'adaptation gap' between what is needed to reduce or avoid impacts and what action is occurring. In this way, insurance is emerging both as part of the call for adaptation and, as with its relation to mitigation, an urgent response to its failures to date. As such, insurance offers a relatively subdued vision of how life is to be secured under climate change. Its sombre tone is accentuated by its continual emphasis on risk and disaster, and its lack of transformational ambition, in contrast to other more creative and justice-oriented approaches to climate change adaptation. It is also subdued because although adaptation is a 'growth area,' expanding under climate change like a new modernisation frontier, it is itself acutely vulnerable to climate change impacts and faces severe, inescapable limits (Mechler et al. 2020). Commercial insurance is a case in point. As Collier et al. (2021, p. 164) explain about catastrophe insurance:

> Discussions of catastrophe insurance are characterised equally by urgent calls to dramatically expand insurance cover for climate-related risks, and warnings that, unless mitigation or adaptation measures are taken, existing insurance arrangements may collapse, and the risks faced by certain populations in certain geographical areas may become uninsurable.

Geoengineering advocates would argue that it should be added to Collier et al.'s list of measures needed to protect the insurance industry from an uninsurable future. In positioning geoengineering as a protector of the insurance industry, geoengineering's narrative about the future not only repeats the refrain of insurance, but also comes to engulf it. The basic insurantial narrative under climate change is *'If* climate change continues, *then* everyone faces great risks... But *what if* they had insurance? This would reduce the impacts.' It moves from warning to solution, demonstrating Modern prudence and ingenuity. Geoengineering repeats the same sequence, but in a way that enrols the insurance industry as an entity at risk, offering it a partial and precarious solution – '*What if geoengineering reduced the impacts?*' – just as insurance offers a partial and precarious solution to others.

Whether the posited solution is provided by geoengineering or other forms of insurance, the very circulation of these '*what if?*' proposals dampens the urgency of '*If this continues, then...*' warnings about the future. We come then to the second reason where insurance and geoengineering, especially, are contested. This is the issue that – counter to insurance's claims of Modern rationality and objective foresight – it is not independent of the future risks it foresees; it helps generate them. To begin with, it can influence people's other risk-reduction behaviours by dangling a 'moral hazard.' The most widespread criticism of insurance is that it ends up substituting for, rather than complementing or encouraging, people's efforts to reduce risk. In the health field, for instance, research suggests that if someone has health insurance, they are less likely to engage in healthy behaviours (Einav & Finkelstein 2018). In climate adaptation, insurance can encourage a reactive, incremental, individualistic approach and hamper more collective, transformational efforts (Wilson et al. 2020).

In the case of geoengineering, opponents argue that more than simply expressing a lack of confidence in greenhouse gas mitigation, the very possibility of geoengineering may undermine mitigation progress by reassuring society (especially large emitters) that regardless of how half-hearted their mitigation efforts are, climate change will be kept manageable. As Parker (2014, p. 7) argues about Solar Radiation Management [SRM] specifically:

> The seductive promise that SRM could be a cheap, quick solution to climate change is one of the greatest concerns about its development. It seems inescapable that at least some people will find that the prospect of a techno-fix for global warming—however imperfect—is cause enough to significantly reduce efforts to curb GHG emissions.

The concern is that geoengineering becomes a self-fulfilling prophecy: demotivating mitigation, worsening climatic change, increasing urgency, and thus bolstering its own appeal as a 'weapon of last resort' (Lederer & Kreuter 2018).

Research into geoengineering options wears two hats here. On the one hand, critics warn that by making geoengineering more technically feasible, research into geoengineering options deepens the moral hazard. On the other hand, advocates present research as a kind of insurance policy itself – one that prepares for the possibility that the moral hazard effect eventuates, climate change worsens, and society turns to the research world seeking ready-made geoengineering solutions or explanations of what to avoid (Parker 2014).

That geoengineering interventions are *themselves* risky is the third main reason they are contested, and a further illustration of how they shape the future. Despite the purported rationalism that underpins its insurantial imaginary, geoengineering in practice is highly speculative, to put it politely. Some initiatives seem more at home in bad science fiction than good science journals. Here, the *'What if...?'* stance on the future is strongly evident. Ideas posited include dumping tonnes of iron into oceans or spreading crushed rock (e.g. mine tailings) over vast areas to absorb

carbon, erecting mechanical trees to suck CO_2 out of the air, or growing and burning real trees, capturing their carbon and sticking it underground (BioEnergy + CCS). More than mere ideas, many of these are
being trialled, formally and informally. While a lack of monitoring and
transparency means the empirical results of many such interventions are
poorly understood, there is clear potential for severely negative impacts
on systems near and far. Even in the testing phase, both SRM and carbon
removal often involve injecting substances of various kinds into the environment, raising the 'problem of permissible pollution' (Hale & Dilling
2011). Many geoengineering options are what international law considers
'ultrahazardous' (Brent 2018) and require never-ending maintenance to
work and to avoid the disruptions to the climate that would result should
large-scale geoengineering suddenly fail (Wong 2014). Social risks are
equally serious. Numerous authors point out the autocratic tendencies of
geoengineering projects and their potential to trigger conflict at various
scales, from individual projects, to regional climate zones to geopolitical
disputes between powerful nations (Dalby 2015; Rabitz 2016; Szerszynski
et al. 2013). SRM, in particular, is read as a form of 'stratospheric imperialism' (Surprise 2020), 'an extreme, expert–elite technocratic intervention
into the global climate system that would serve to further concentrate contemporary forms of political and economic power' (Stephens & Surprise
2020, p. 2; Schneider 2019).

'Rogue geoengineering' could affect the insurance sector badly, given
the latter's reliance on knowing the risk landscape. For, if geoengineering
were to change weather patterns, as it wants to but may do in unknown and
unintended ways, the ability of insurers and others to calculate and insure
against specific climatic risks would be badly diminished. That said, for
commercial insurance and its '*what if?*' creative responses, where there is
risk there is opportunity. Thus, alongside other compensation and liability
mechanisms, insurance is being introduced as an antidote to some of geoengineering's possible side-effects – or at least, investors' liability for such
side-effects (Packard 2018). In particular, parametric insurance is being
proposed as means to govern SRM, building on the use of such insurance
across large regions as a climate adaptation mechanism (Horton et al. 2020).
In a given region, a parametric insurance scheme would be underwritten by
those implementing geoengineering, who would pay out others if certain
climate indices are reached (Horton & Keith 2019). It goes without saying
that the feasibility of such a scheme is highly uncertain. The point is that
rather than insurance companies rejecting geoengineering as delusional,
they are exploring the deployment of geoengineering as a new insurance
market, further illustrating the resilience and tumbling interconnections
of commercial insurance and geoengineering, as both are tapped along by
their common insurantial logic, nested '*what if?*' propositions, shared capitalist orientation and scrambling efforts to keep up with their and others'
cascading effects in the world.

Turning to the terrestrial

Many of the geoengineering interventions mentioned above are up in the air figuratively as well as physically. In contrast, terrestrial carbon sequestration is down to earth, if far from tried and true. 'Technological' versions of carbon removal, such as Carbon Capture and Storage, are often terrestrial in the double sense of capturing CO_2 from the combustion of once-subterranean fossil carbon (coal, oil, or gas) and related industrial processes, then extracting, liquifying, transporting, and injecting the liquid back underground, often into empty oil and gas reservoirs. Initially invented by the oil industry to extract remnant oil from reservoirs, CCS is a highly energy-intensive process that, like all carbon removal methods, relies on the assurance that it will never leak the CO_2 into the atmosphere. It is not a confident promise. The entire value chain and production cycle is replete with leakage risks (Guthrie & Kirrane 2017). For the insurance industry, though, this opens up yet another risk frontier and market. As Swiss Re (2020) beckons: 'By 2050, billions of tons of CO_2 will need to be stored: the front-runners among insurers will profit from the experience gathered over the next decade.' Insurance products for CCS are indeed emerging.

At the same time, the possibility of catastrophic CO_2 loss (and related human poisonings and impacts on the climate) threatens to limit the new market. There is a non-negligible likelihood of mass CO_2 release caused by seismic activity, especially given that such instability seems to be exacerbated by both climate change and the injection of liquids into geological formations, as CCS does (Buis 2019; Masih 2018). Whether or not these feedbacks are being taken into account, the risks posed by a CCS project are being judged uninsurable in some cases (Guthrie & Kirrane 2017), reflecting again insurance's constrained and two-sided relation with climate change-related risks.

The most terrestrial of terrestrial carbon sequestration techniques is that which positions *soil* as the key infrastructure. Rather than imagined as a mere storage vessel, soil is imagined here as a world in miniature, including busy factories of microbial, insect, plant, and fungi workers actively pulling carbon in from the air and securing it in myriad forms (Krzywoszynska 2020). Adding to the attractiveness of soil carbon sequestration is the fact that much of it is being pursued on farmland (e.g. Australia's *Carbon Farming Initiative*), enrolling the agricultural sector into climate change action in a refreshingly positive way, and giving geoengineering agrarian appeal (Kearnes & Rickards 2020). For farmers where 'carbon farming' schemes exist, carbon sequestration offers an income diversification option and thus some 'insurance' to farm businesses against poor production profitability. More significantly, it offers co-benefits, notably improvements in soil health and climate resilience, particularly when pursued as part of a broader 'regenerative' agriculture approach (Kearnes & Rickards 2020). As

Droste et al. (2020, p. 1) put it: 'soil carbon provides farmers with a natural insurance against climate change through a gain in yield stability and more resilient production.' Soil is positioned here as not only a form of infrastructure, but a form of *natural insurance,* underlining how – consistent with wider capitalist processes – nature is being enrolled into human climate responses as not just threat but tool.

Commercial insurance features within these agricultural soil carbon efforts in characteristically heterogeneous ways. Insurance companies and banks encourage farmers to better adapt their farms to harsher climates and, like others, are beginning to explicitly advocate better soil management as part of this, given it helps protect them against insurance payouts and loan defaults (Kane et al. 2021). At the same time, there is a tension between commercial and 'natural insurance,' with farmers protected by the latter (i.e. healthy soils) showing less interest in commercial products such as crop insurance (Jørgensen et al. 2020). Nevertheless, formal carbon sequestration schemes represent an insurance opportunity, and insurers now offer insurance to those taking on soil carbon sequestration contracts, given that (as with other forms of geoengineering) there are uncertainties about the permanence of soil's carbon removal services, particularly given more frequent and severe disturbances to soil under climate change in the form of droughts, floods, wind, and fires[2].

A less recognised but highly influential way that the insurance industry is involved in soil-based geoengineering is as a major owner and even manager of farmland. For instance, Canada's largest insurer, Manulife Investment Management Company, invests funds for a large number of retirement funds. It owns Hancock Natural Resource Group (HNRG), which in turn owns farmland across North America and Australia, actively managing about half of it (Fairbairn 2020). Hancock was one of the first insurance-based asset managers to turn to farmland and help make it into the sought-after 'asset class' it is today (Fairbairn 2020; Ouma 2020). Key to this conversion was farmland's appeal as a secure investment relative to financial stocks, giving it an insurance-like quality for investors (Fairbairn 2020). Also important were improvements in managing the downside risks inherent to agriculture, especially under climate change. While institutional farmland investors rarely intend to own the land into perpetuity in the way many farm families do, as indicated above they are increasingly recognising the importance of good soil management, not the least to meet social and legal expectations. HNRG, for instance, now encourages its farm managers to practice (some) regenerative farming techniques as a way of delivering its investors 'enhanced farmland value' and helping the company demonstrate responsible investing (HNRG 2020, p. 5).

Insurers, other investors, 'carbon farming aggregators,' and other consultants are all working to assemble international soil carbon markets, pulling soil out of the dull light of the rustic into the bright light of modern

Capitalism. In linking the molecular with the Global in this way, and seizing upon carbon and farmland as new assets, they are perpetuating the standard model of capitalist expansion and its never-ending refrain of *'what if?'*. It is worth considering, however, whether insurers' acute interest in not only expansion and gain, but limits and threats, gives them some inkling of the unsustainability of this model and the profound insecurities that now characterise human futures. Like geoengineering, insurance is two-faced, both celebrating human capacities to work with nature in the form of solar radiation and geochemical flows, while emphasising the likelihood of rupture and harm of the sort the planet's recalcitrant inhuman agency is increasingly demonstrating through blows against insurance itself. Even at the farm scale, insurers need (like and as farmers) to be alert to the potential for 'soil refusal' (Tironi 2020) – situations in which soil fails to perform as expected or needed, regardless of whether the intent was crop yield, carbon sequestration, or caring entanglement. A growing reality under climate change, soil refusal includes the possible rejection of vast amounts of carbon out of the terrestrial sphere, rendering it less lively and repudiating the modernist dream that soil will provide us with a reliable escape from the streams of greenhouse gases flowing unabated behind capitalism's ascent. Insurance's focus on *'If..., then...'* warnings about the future reminds us of the limits to merely romanticising soil or trying to capitalise increasingly violent Anthropocene feedbacks. It forces us to ask: 'how are we to think about soil-human relationality when soil does not render itself relational?' (Tironi 2020, p. 184). 'How are we to act if the territory itself begins to participate' in the world (Latour 2018, p. 41)? How are we to assist others when our own very future existence is in question? In its commercial form, insurance offers only an orthodox modernist *'what if?'* solution. But what if the broader insurantial imaginary – or at least its undercurrents of reciprocity and solidarity – could help reorient us to another future, another narrative mode? An *'if only...'* mode of storying that dismisses the escapist fantasy of the Out-of-This-World utopia and asks instead what it would mean *if only* we knew how to relate to or at least live with this new Earth. Perhaps a new insurantial imaginary, grounded in 'the elementality of soil' (Tironi et al. 2020, p. 27), could break its modernist habits, call out the negations of delusional forms of geoengineering, take up the climate justice fight, and help turn us towards 'the Terrestrial' (Latour 2018), compelling us to negotiate the violence and promise of the inseparability of earth, Earth, and us earthlings.

Notes

1 This typology is discussed by the late science fiction writer Octavia Butler, who attributes its origin to Robert A. Heinlein. See Butler, Octavia. "'Devil Girl From Mars': Why I Write Science Fiction." Lecture, MIT, February 19, 1998. http://web.mit.edu/m-i-t/articles/butler_talk_index.html. Thank you to Mathias Thaler for pointing me to Octavia's lecture.

2 For example, Carbon Farming Australia is one of many consulting firms in Australia that aggregate together individual farms' carbon sequestration contracts, liaise with the funder (usually the federal government), and manage the contracts for the farmers, including arranging insurance for them. https://carbonfarmersofaustralia.com.au/carbon-trading/getting-started-with-the-carbon-farming-initiative/

References

Anshelm, J & Hansson, A 2014, 'The last chance to save the planet? An analysis of the geoengineering advocacy discourse in the public debate', *Environmental Humanities*, vol. 5, pp. 101–123.

Anshelm, J & Hansson, A 2016, 'Has the grand idea of geoengineering as Plan B run out of steam?', *The Anthropocene Review*, vol. 3, pp. 64–74.

Booth, K & Tranter, B 2018, 'When disaster strikes: Under-insurance in Australian households', *Urban Studies*, vol. 55, pp. 3135–3150.

Brent, K 2018, 'Solar radiation management geoengineering and strict liability for ultrahazardous activities', In Jefferies, C.S.G, Craik, N, Seck, S.L, Stephens, T (Eds.) *Global Environmental Change and Innovation in International Law*, Cambridge University Press, Cambridge, pp. 161–179.

Buis, A 2019, 'Can climate affect earthquakes, or is the connection shaky?', *NASA Global Climate Change, Vital Signs*, viewed 29 October 2019, <https://climate.nasa.gov/news/2926/can-climate-affect-earthquakes-or-are-the-connections-shaky/>.

Collier, S.J, Elliott, R & Lehtonen, T-K 2021, 'Climate change and insurance', *Economy and Society*, vol. 50, pp. 158–172.

Dalby, S 2015, 'Geoengineering: The next era of geopolitics?', *Geography Compass*, vol. 9, pp. 190–201.

Dannenberg, A & Zitzelsberger, S 2019, 'Climate experts' views on geoengineering depend on their beliefs about climate change impacts', *Nature Climate Change*, vol. 9, pp. 769–775.

Droste, N, May, W, Clough, Y, Börjesson, G, Brady, M.V & Hedlund, K 2020, 'Soil carbon insures arable crop production against increasing adverse weather due to climate change', *Environmental Research Letters*, vol. 15, no. 12, p. 124034.

Einav, L & Finkelstein, A 2018, 'Moral hazard in health insurance: What we know and how we know it', *Journal of the European Economic Association*, vol. 16, pp. 957–982.

Ewald, F 1991, 'Insurance and risk', In Burchell, G, Gordon, C, Miller, P (Eds.) *The Foucault Effect: Studies in Governmentality*, University of Chicago Press, Chicago.

Ewald, F 2019, 'The values of insurance', *Grey Room*, pp. 120–145.

Fairbairn, M 2020, 'Fields of gold', *Fields of Gold*. Cornell University Press, Ithaca, NY.

Fragniere, A & Gardiner, S.M 2016, 'Why geoengineering is not 'Plan B''. In Preston, CJ (Ed.) *Climate Justice and Geoengineering: Ethics and Policy in the Atmospheric Anthropocene*, Rowman & Littlefield International, London, pp. 15–32.

Guthrie, D & Kirrane, A 2017, 'Insurance issues for carbon capture and storage', *HWL Ebsworth Lawyers*, <https://hwlebsworth.com.au/insurance-issues-for-carbon-capture-and-storage/>.

Hale, B & Dilling, L 2011, 'Geoengineering, ocean fertilization, and the problem of permissible pollution', *Science, Technology & Human Values*, vol. 36, pp. 190–212.

HNRG 2020, 'The promise of regenerative agriculture: Environmental and conservation benefits of conservation tillage and cover cropping', *Hancock Natural Resource Group*, <https://hancockagriculture.com/wp-content/uploads/sites/3/Conservation-Tillage-and-Cover-Crops.pdf>.

Horton, J.B & Keith, D.W 2019, 'Multilateral parametric climate risk insurance: A tool to facilitate agreement about deployment of solar geoengineering?', *Climate Policy*, vol. 19, pp. 820–826.

Horton, J.B, Lefale, P & Keith, D 2020, 'Parametric insurance for solar geoengineering: Insights from the Pacific catastrophe risk assessment and financing initiative', *Global Policy*, vol. 12, Suppl. 1, pp. 97–107.

Jørgensen, S.L, Termansen, M & Pascual, U 2020, 'Natural insurance as condition for market insurance: Climate change adaptation in agriculture', *Ecological Economics*, vol. 169, p. 106489.

Kane, D.A, Bradford, M.A, Fuller, E, Oldfield, E.E & Wood, S.A 2021, 'Soil organic matter protects US maize yields and lowers crop insurance payouts under drought', *Environmental Research Letters*, vol. 16, p. 044018.

Kearnes, M & Rickards, L 2020, 'Knowing earth, knowing soil: Epistemological work and the political aesthetics of regenerative agriculture', In Salazar, J, Kearnes, M, Granjou, C, Krzywoszynska, A, Tironi, M (Eds.) *Thinking with Soils: Social Theory and Material Politics*, Bloomsbury Publishing, London, pp. 71–88.

Kiehl, J.T 2006, 'Geoengineering climate change: Treating the symptom over the cause?' *Climatic Change*, vol. 77, pp. 227–228.

Krzywoszynska, A 2020, 'Nonhuman labor and the making of resources: Making soils a resource through microbial labor', *Environmental Humanities*, vol. 12, pp. 227–249.

Latour, B 2018, *Down to Earth: Politics in the New Climatic Regime*, John Wiley & Sons, New Jersey.

Lederer, M & Kreuter, J 2018, 'Organising the unthinkable in times of crises: Will climate engineering become the weapon of last resort in the Anthropocene?', *Organization*, vol. 25, pp. 472–490.

Markusson, N, Ginn, F, Singh Ghaleigh, N & Scott, V 2014 "In case of emergency press here': Framing geoengineering as a response to dangerous climate change', *Wiley Interdisciplinary Reviews: Climate Change*, vol. 5, pp. 281–290.

Masih, A 2018, 'An enhanced seismic activity observed due to climate change: Preliminary results from Alaska', *IOP Conference Series: Earth and Environmental Science*, IOP Publishing.

Masson-Delmotte, V et al. 2018, 'Global warming of 1.5°C', IPCC, Geneva. https://www.ipcc.ch/sr15/.

McCracken, M 2009, *Geoengineering: A Possible Climate Change Insurance Policy*, WorldBank, <https://blogs.worldbank.org/climatechange/geoengineering-possible-climate-change-insurance-policy>.

Mechler, R, Singh, C, Ebi, K, Djalante, R, Thomas, A, James, R, Tschakert, P, Wewerinke-Singh, M, Schinko, T & Ley, D 2020, 'Loss and damage and limits to adaptation: Recent IPCC insights and implications for climate science and policy', *Sustainability Science*, vol. 15, pp. 1245–1251.

Ouma, S 2020, 'This can('t) be an asset class: The world of money management, 'society', and the contested morality of farmland investments', *Environment and Planning A: Economy and Space*, vol. 52, pp. 66–87.

Packard, L 2018, 'Designing an international liability regime to compensate victims of solar radiation management', *Environmental Claims Journal*, vol. 30, pp. 71–86.

Parker, A 2014, 'Governing solar geoengineering research as it leaves the laboratory', *Philosophical Transactions of the Royal Society A: Mathematical, Physical and Engineering Sciences*, vol. 372, p. 20140173.

Rabitz, F 2016, 'Going rogue? Scenarios for unilateral geoengineering', *Futures*, vol. 84, pp. 98–107.

Schneider, B 2017, 'The future face of the earth: The visual semantics of the future in climate change imagery of the IPCC', In Heymann, M, Gramelsberger, G, Mahony, M (Eds.) *Cultures of Prediction in Atmospheric and Climate Science: Epistemic and Cultural Shifts in Computer-Based Modelling and Simulation*, Routledge, London, pp. 231–251.

Schneider, L 2019, 'Fixing the climate? How geoengineering threatens to undermine the SDGs and climate justice', *Development*, vol. 62, pp. 29–36.

Stephens, J.C & Surprise, K 2020, 'The hidden injustices of advancing solar geoengineering research', *Global Sustainability*, vol. 3, pp. 1–6.

Surprise, K 2020, 'Stratospheric imperialism: Liberalism, (eco)modernization, and ideologies of solar geoengineering research', *Environment and Planning E: Nature and Space*, vol. 3, pp. 141–163.

Swiss Re 2020, 'Special feature: Locking it up – carbon removal and insurance', *SONAR 2020: New Emerging Risk Insights*, <https://www.swissre.com/institute/research/sonar/sonar2020/sonar2020-carbon-removal-insurance.html>.

Szerszynski, B, Kearnes, M, Macnaghten, P, Owen, R & Stilgoe, J 2013, 'Why solar radiation management geoengineering and democracy won't mix', *Environment and Planning A*, vol. 45, pp. 2809–2816.

Tironi, M 2020, 'Soil refusal: Thinking earthly matters as radical alterity', In Salazar, J, Kearnes, M, Granjou, C, Krzywoszynska, A, Tironi, M (Eds.) *Thinking with Soils: Social Theory and Material Politics*, Bloomsbury Publishing, London, pp. 175–190.

Tironi, M, Kearnes, M, Krzywoszynska, A, Granjou, C & Salazar, J.F 2020, 'Soil theories: Relational, decolonial, inhuman', In Salazar, J, Kearnes, M, Granjou, C, Krzywoszynska, A, Tironi, M (Eds.) *Thinking with Soils: Material Politics and Social Theory*, Bloomsbury Publishing, London, pp. 15–38.

Wilson, R.S, Herziger, A, Hamilton, M & Brooks, J.S 2020, 'From incremental to transformative adaptation in individual responses to climate-exacerbated hazards', *Nature Climate Change*, vol. 10, pp. 200–208.

Wong, P-H 2014, 'Maintenance required: The ethics of geoengineering and post-implementation scenarios', *Ethics, Policy & Environment*, vol. 17, pp. 186–191.

3 Indexing the soil

Olli Hasu and Turo-Kimmo Lehtonen

Introduction: A close reading of a programmatic document by the World Bank

How is the future of the soil financialised? In this chapter we look into a technology, index insurance, that promises to make the productivity of the earth and its uncertainties manageable. Index insurance is a tool that in the build-up phase of its design takes into account a wide variety of heterogeneous variables, such as the broad environmental system in an area, its historical weather conditions and their changes, and the social conditions in which the soil is processed; yet, it ends up abstracting most of these variables in favour of a streamlined economic model that concentrates on the likelihood of payouts. What is thus produced is a dynamic tool for translating local agricultural conditions so they can be assessed from the point of view of global financial markets that can subsequently intervene in local processes from afar.

Our study contributes to research about the financialisation of natural environments, which refers to the assetisation of ecological metrics, such as extreme weather event and carbon emission data, and to the growing influence of finance in guiding political governance (Chiapello 2020; Goodman & Anderson 2020; Langley 2020; Ouma et al. 2018). We focus specifically on an index insurance risk model that uses environmental data to design a social infrastructure for governing weather-related hazards. The rules and principles established in the modelling process constitute a technical methodology for perceiving risks in ecological systems. Through this methodology, index insurance not merely represents natural phenomena but rather generates governable environments as an assemblage of four elements: soil, information-technology, financial risk modelling, and social coordination. In this way, index insurance provides a case study on how finance mediates ecological environments into socioeconomic constructions.

Weather-related forms of insurance have in recent years gained growing attention among social scientists. Indeed, there already exists a relatively large body of research discussing how climate change adaptation and mitigation are pursued through different kinds of insurance instruments (Angeli

DOI: 10.4324/9781003157571-4

Aguiton 2019; 2020; Bridge et al. 2020; Christophers et al. 2020; Collier & Cox 2021; Collier et al. 2021; Elliott 2021; Gray 2021; Grove 2021; Johnson 2013; 2021; Lehtonen 2017; Lucas & Booth 2020; Taylor 2020). These studies make it evident how widely shared, among both public and private actors, is the understanding that insurance technology is an obligatory passage point for translating large-scale environmental hazards into actionable issues. Research shows that, in fact, relevant financial technologies come in many forms, that their use can be highly context specific, and that they can be contested for good reasons. Nevertheless, what remains constant across the field is the perception that the changing risks generated by climate change create threats and opportunities for the industry; climate change is at the core of present-day discussions on insurance and weather-related catastrophes.

This chapter is based on a close reading of the World Bank Global Index Insurance Facility (GIIF) document: *Risk Modeling for Appraising Named Peril Index Insurance Products: A Guide for Practitioners* (*RM* below; Mapfumo et al. 2017). The project articulates its general aims as follows: 'GIIF's objectives are to provide access to financing for the vulnerable; to strengthen the financial resilience of the poor against the impact of climate change and natural disasters, and to sustain food production for local communities and larger markets.' Within GIIF, *RM* has been used for workshops and course material, such as *Emerging Guidelines for Underwriting and Portfolio Management*. We concentrate our analysis on *RM* because of the document's programmatic and authoritative nature. What makes the text especially interesting is how it provides normative guidelines for putting together and employing index insurance and presents arguments about how to use – and not use – the multiplicity of environmental data to design mechanisms of socioeconomic coordination.

We read *RM* to examine three themes that it unveils. First, we scrutinise the practical means through which the soil is transformed into the index, an abstract object of calculation. The soil itself is a complex entity that is comprised of myriad living beings, processes, and interactions with weather conditions and human intervention. The index performs a selection of the soil's elements in a process mediated by satellite technologies, information infrastructures, and forms of modelling.

Second, we analyse why and how the index is used. It is revealed to be a technology that transforms local uncertainties regarding the soil, weather dynamics, and yield into financial objects. Index insurance creates a specific kind of orientation to caring for future uncertainties. It provides a distribution channel for financial services while also creating a method for formalising expectations about environmental risks as economic factors. In other words, index insurance 'objectifies' weather-related catastrophes (Lehtonen 2017): their past occurrences are taken into account for defining the likelihood of future hazards, and the calculation of past and potential future losses in terms of monetary value render these catastrophes into economic objects, 'risks.' As weather-related catastrophes are analysed in terms of precise monetary values, the risks involved can be treated on the same objectified level as that of all kinds of other financial instruments. In other words, as risks that are related to agricultural practices and the caring for soil, they become comparable to all other financial cost–benefit analyses and turn into entities that can be traded in international markets and that financial actors can invest in. This chapter considers how the instrument works as part of a programme that connects local economies to external resources, representing a systematic strategy to integrate the soil into the coordination of global financial markets.

Third, the work on the two previous themes has led us to a surprising finding as regards the contents of *RM:* although index insurance is much advertised by the World Bank and GIIF as an efficient tool for engaging with financialised climate change mitigation, in *RM*, a lengthy document of more than 300 pages, the term of 'climate change' occurs only once; moreover, as will be detailed below, the instrument is not intended to take into account *risks that change*, thus effectively precluding from its scope of intervention the very idea of climate *change*. Thus, the instrument is revealed to be a means of objectifying weather-related risks as something that the financial infrastructure can intervene in and profit from, even if the high hopes of 'climate change mitigation' are completely sidestepped.

While examining these themes, we obviously rely on the recent social scientific literature on weather-related insurance technologies and reinsurance. As we focus on a tool developed under the auspices of the World Bank, our research draws especially on the work of Leigh Johnson and colleagues (Johnson 2013; Johnson et al. 2019). In a recent article, she describes 15 years of index insurance development and experimentation that has sought to expand insurance coverage to the poorest regions of the world in order to build resilience against climate change risks (Johnson 2021). The chronicling of multiple programmes reveals a largely failed project that is suffering from both low demand and significant problems in product design. Analysing institutional composition, governmental goal articulation, and strategies for correcting the instruments' apparent flaws, Johnson identifies a change in the development of index insurance products, which are shifting from microfinance towards meso- and macro-level instruments. Her analysis underscores the political economy of climate risk management. Index insurance products are designed for areas where weak institutional capabilities make preparing for climate change-caused shocks difficult. Therefore, even an unreliable risk technology can be received with enthusiasm in regions defined by their vulnerable position in the global economy (see also Grove 2021).

In contrast to Johnson's synthesising interpretations of the uses of index insurance in developing contexts, in this text we concentrate on analysing the design of the instrument, as represented in the core document *RM*. In doing so, our aim is to tease out the technological underpinnings of the index insurance endeavour. We want to dig deeply into understanding how the insurantial perception of soil is made up, or – to paraphrase the famous text by James Scott (1998) – what 'seeing like an index insurance' entails.

The structure of the chapter is as follows. First, we explicate the way in which the index insurance is assembled according to the *RM* document. Then, we explain how at the core of the instrument's operations is mapping a region and modelling differences between areas in that region. This leads us to the next two sections. In the first, we describe *RM*'s different ways of spatialising time in the modelling work, and then, in the second, go deeper into how the advertised forms of modelling in fact completely exclude environmental change. Finally, we conclude by highlighting our surprising main finding: although index insurance is promoted by GIIF as a tool that helps deal with climate change, the programmatic guide that it proposes for practical uses, *RM*, completely bypasses this theme area and narrows stakeholders' attention to the technical calculations of local payout ratios.

The objectives of the index

In *RM*, the World Bank renders index insurance comprehensible for a variety of stakeholders and explains how it can be used to manage the agricultural economy in developing countries. After a general introduction to the purpose of index insurance, *RM* consists of two substantial parts: the first describes and advertises the decision tools available for insurance managers, and the second explains how probabilistic modelling works for insurance analysts. Altogether, the document is 315 pages long. In the very first pages of *RM*, the authors discuss who they see as its ideal audience and for whom the detailed exposition of the advertised risk management tool will be useful. The primary readership will be composed by 'managers and actuarial analysts of insurance companies in developing countries' but also by more local intermediaries through whom small farmers and their service providers can be reached: the '[f]armer organizations, financial institutions, and agriculture value chain actors and investors evaluating the potential benefits and risks of index insurance policies' (*RM*: 1). In addition, the authors see *RM* as a useful document both to regulators involved in 'assessing insurance products for client value and consumer protection purposes' and students 'interested in quantitative risk analysis and probabilistic modeling' (*RM*: 1). While expecting to persuade such a broad constituency to develop active interest in the tool, the authors state even broader aims in the Foreword; they hope that the instrument will not only advance 'financial inclusion' but also increase investment in 'smart agricultural technologies' (*RM*: xvii). Behind all this is the idea that the agricultural sector in developing countries deserves more attention from global financiers. The 'unserved market segment' of small farmers can form an 'attractive customer base' for insurers and, consequently, the guide is presented as a tool that helps 'emerging market insurers' to 'penetrate new market segments' (*RM*: xvii; see also *RM*: 83).

Historically, low premium volumes and expensive operating costs have created a critical obstacle for insurance market expansion in the Global South, where indemnity insurance is typically regarded as financially untenable. In the aftermath of natural disasters, infrastructure damage makes field assessments difficult and slow. In this respect, the advantage of index insurance lies in its cost efficiency. Instead of examining the damage suffered item by item, as traditional forms of insurance do, index insurance objectifies environmental risks as geographically standardised phenomena. This is the reason it can operate automatically and symmetrically in relation to each policyholder within a specified area.

Assembling the index

It is important to understand that index insurance does not exist as a ready-made tool that travels easily and can be readily applied to different environmental settings. Rather, as an instrument, it comes into existence through an intricate design process for a specific purpose; it requires that various actors and institutions – from smallholder farmers to professionals in data analysis and finance – come together to form a network that is able to model predictively and yield the intermittent weight of environmental shocks. Additionally, index insurance is commonly bundled with other financial services, such as credit.

As detailed by *RM* (11–13), such a network includes, first, the *product design* team, a separate entity often consisting of international experts with special skills required for developing the instrument. Second, the *data processing team* makes automated real-time data-based claim processing possible. Information sources can include weather stations, remote sensing technology, and satellites. Third, *data providers* are public or private institutions that provide both historical and real-time information for pricing and automatic claim processing. Fourth, the network depends on the activity of a *regulator* that sets norms and approves the issuing of a product. Fifth, the *insurer* then issues the product, collects premiums, reinsures part of the portfolio, and handles any claims that arise. Sixth, the *reinsurer* underwrites some or all of the insured risks. Because index insurance protects against systemic risks, a large part of the insurer's portfolio should be reinsured on the global financial markets. Seventh, the *insured party* carries out the transaction to transfer risks to the financial market. Eighth, and finally, the *policyholder* is often not the same as the insured party. For example, in the case of smallholder farmers, the policyholder is usually an aggregator that makes the issuing of policies more attractive to insurers; this role can be played by, for instance, a cooperative, a microfinance institution, or a commercial bank.

In the rest of this section we analyse the key moments of the work that lead up to the finished index. These include: how data is gathered; how coverage is determined and how the payouts are structured; the importance of creating maps; and the question concerning what in the business is called 'basis risk.'

Data sources

In the design process, a variety of sources are used to assemble information. Typically, this will include historical hazard data, inventory damage figures, and local expert knowledge from specialists such as agronomists, hydrologists, and seismologists. Where historical quantitative data is lacking, anecdotal accounts are used: 'the product design team relies on farmers' recollections and information from local experts as well as government and international sources to categorize the level of crop damage caused by the named peril in each year and geographical area' (*RM*: 34).

Determining the structure of coverage and payments

Index insurance transforms all these pieces of information and streams of visual or quantitative data from satellites and weather stations into a financial model that makes payouts when a specified threshold is reached in the monitored data. The payout triggers are defined as a percentage of the sum insured. For example, a policy can be designed so that the insured will be indemnified when a region's cumulative rainfall for the policy period is under 100 millimetres, with each millimetre below the trigger equalling 2% of the sum insured; thus, 100% of the sum insured is paid out when the cumulative rainfall is less than 50 millimetres.

The design process begins with constructing a base index that provides full coverage on the modelled risk events. However, to produce a marketable insurance instrument, it does not suffice to establish the environmental likelihoods in a given area. For potential policyholders, the high coverage of the base index is often too expensive. Therefore, the next step in the process is to redesign the index so that it provides less coverage but is cheaper and better fits the economic interests between local farmers and the insurer. Thus, as described by the document (*RM:* 17), in practice index insurance will usually be saleable only as a product that *underinsures* the relevant risks.

Mapping

Risk categorisation for the instrument's purposes is achieved in geographic terms. The levels of expected average damage are estimated by organising a region into specified areas with determined risk profiles. The idea is that when a payout is triggered for an area, all insured farmers within it receive the same amount of compensation; no differentiation between policyholders is made. This is the reason why index insurance products do not require individualised damage evaluations to process payouts. The other side of the coin is that *mapping* becomes the crucial activity for making index insurance feasible in economic terms (*RM:* 154). Mapping, for its part, gains its full effect only through the way in which it is linked with modelling.

To objectify environmental and weather-related risks, a map of spatially distributed risk factors is created, and the assemblage of these factors is treated as a proxy for events that cause damages for farmers. Because the data is processed by third-party providers, the objectivity of index-based risk modelling is institutionally guaranteed and thus the insurance policies can be transferred to global financial markets. From an insurer's point of view, the area-based perception of risks has the important benefit of eliminating moral hazard in the contract, as it is impossible for policyholders to affect the likelihood of payouts with their own behaviour (*RM:* 9–10).

On top of the map representing soil and weather risk patterns, a layer of pricing models is added to define how the spatially standardised risk events can be insured, thus providing the socioeconomic logic for the process (*RM*: 10). The end result is an index that should be able to represent financially homogenous risk events that affect all policyholders uniformly within a specified geographic area:

> Based on the agreed-on inputs, the actuarial analyst produces equitable premiums for each geographical area. [...] In this case, the goal of the analysis is not to find one overall premium rate that can be applied to the total portfolio of geographical areas, but to find the equitable premium for each area that takes into account each area's specific characteristics and risks. It is important to note that the equitable premium is for the area, not for individual insured units.
>
> (RM: 62)

Basis risk

However, the area-based standardisation of risk information is simultaneously the main modelling-related problem that has thus far appeared unresolvable for index insurance projects. Indeed, Johnson (2021) argues that a central reason for the failures of index insurance programmes is the *basis risk* that plagues the product design. Basis risk refers to the difference between risk events represented by the index and the actual losses experienced by the policyholders. In other words, if the payout trigger levels defined by an index insurance product do not accurately correspond with the actual damage that the instrument models, there will be situations where policyholders have paid their premiums, yet suffer losses caused by the very risk event that the product is supposed to cover. According to Johnson, this raises the question of whether index insurance can fulfil its assumed potential as a risk technology (Johnson 2021). However, it is significant that, according to *RM*, such situations are simply *inevitable:* 'It is important to note that there will be situations in which an insured party experiences a loss attributable to a hazard event but does not receive a payout' (*RM*: 10). Yet the guide elaborates on the theme and claims that this, in fact, is technically not a question of 'basis risk' because index insurance only makes payouts for the risk events defined by the coverage level in the policy. As explained above, in practice, the authors of *RM* think it would be difficult to sell index insurance that would cover the base index and that thus would not imply underinsurance.

Multiple topologies of temporality

The soil on which smallholder farmers live is constituted by complex ecological processes and shaped by changing weather conditions that cause uncertainty. Governing such uncertainty has always been part of agricultural practice and skill. However, commodified risk management brings a new layer to how this is done. In order to successfully financialise the relation to weather-related risks, the unknowable future must be made controllable through a mapping process. The *durée* of the soil is objectified, or to put this in Henri Bergson's (1896) terms, time is rendered *spatial*. Yet, this objectification comes in many forms, not just one. Different ways of conceiving and simultaneously spatialising time interact in the development of the insurance index tool. Therefore, taking into account the observation that time is both spatialised and objectified in multiple forms, it is not out of place to claim that there are different 'topologies' of temporality evident in the design of index insurance.

First, in the early stages of the index insurance design process, the history of the region at which the product will be aimed is mapped (*RM:* 29–30). What kind of variance can be seen? What about disruptions to regularities? Such information is in the background of the product. Yet, if the calculation of probabilities takes into account past events as discrete variables and no attention is paid to the temporal dynamics of their occurrence (for example, by putting more weight on more recent events), time is neutralised and spatialised into a homogeneous field.

Second, the authors acknowledge that regularities could change and that environmental conditions might vary over periodic cycles, if not be fundamentally transformed in a relatively short period, as is the case with regions heavily affected by climate change. However, the term 'climate change' appears only once (*RM:* 125) in the more than 300 pages of the entire document. Somewhat surprisingly, according to *RM*, well-developed index insurance systematically bypasses the view that risks change:

> Weather, and therefore the indexes used in a weather-based index insurance product, may go through multiyear cycles of, for example, dry and wet years. Dry years may be followed by more dry years, and vice versa. Such temporal relationships are not taken into account in the model. The model assumes that any data for the past 30 years are predictive, and more recent data are not more predictive than data from 25 to 30 years ago.
>
> (*RM:* 269)

Thus, RM approaches the soil's dynamics primarily by means of probabilistic modelling where temporality is considered only from the perspective of a flattened time horizon that does not advance. The guide stresses simple and efficient ways of controlling information, whereby for modelling purposes, temporality is primarily treated as a spatialised category.

Third, the situation is slightly complicated by the fact that *RM* recommends using one-year time frames for modelling risks: 'When estimating metrics such as the capital required or the probability of ruin, the models only consider these risks over a one-year horizon' (*RM:* 97). Practically, this implies that the model will take into account incremental change; every year, the previous year's data will be added to earlier data and can thus redirect the model's values, if ever so slightly.

Fourth, while long-period prediction is left out of the modelling, the guide still recommends that actuaries do reflect on scenarios stretching from three to five years to reach a better understanding of the product's likely performance (*RM:* 97). In other words, although the *model* is seen to function best if kept simple and temporal dynamics are left out, its *users* are still advised to retain a broader prudential view in which the model is not their sole source of information.

Fifth, another time frame is given by the global financial markets within which index insurance operates (*RM:* 24). The renewal period of contracts takes place yearly (Jarzabkowski et al. 2015). Prices will go up and down in correlation to other fields where (re)insurers are active and face risk events in a wide variety of business sectors and in all four corners of the globe. Thus, broader financial considerations can profoundly affect the price range in which index insurance operates; these dynamics constitute a timescape of its own that will affect index insurance.

Whichever way temporality is objectified for the purposes of index insurance, it is significant that *RM* does not deem it possible to model temporal change efficiently. The uncertainty included in the modelling of historical data is controlled on the basis that 'future patterns will be similar to those in the past'; in other words, there is no aspiration to 'account for possible changes in the systems themselves over time' (*RM:* 125–6). Such a drastic reduction of the information included has important consequences. Although the ecological environment is taken into consideration in the early build-up of the model, the guide's choice is to assume that the probability of risk events does not alter in the future; the world is perceived as governed by systemic stability. This results in a situation where index insurance in the form advanced by *RM* is *not useful for modelling the impact of climate change*.

Modelling payout ratios

The surprising choice of leaving out temporal dynamics has as its background the aim of making the model as simple and elegant and thus as easily operable as possible. In the guide, a central principle for evaluating the design of insurance products concerns the relation between the complexity of the models used and their intended purpose. An increase in complexity tends to lead to higher resource costs and less predictable performance of the instrument.

RM includes a didactic section in which the authors lay out a theoretical framework for justifying how a system is objectified in the design of index insurance. They explain the thinking behind probabilistic modelling choices and detail how models helpfully simplify reality and serve as tools that fulfil context-specific goals. The guide concentrates on examining systems operationally; that is, as defined on the basis of how they work rather than what they are. The emphasis on operationality is elaborated further in defining the hierarchy of different models that comprise the totality of an index insurance product. Index insurance development uses several submodels for processing economic and ecological data, each of which has additional models defining parameter values. In the formal hierarchy of index insurance design, payout ratio modelling is at the top, while indices for environmental risk data, such as drought frequency and drought severity, are situated as submodels (*RM:* 102). Importantly, this multi-layered apparatus is too complex for calculating definitive values. Instead, index insurance relies on *probability simulations* that generate value approximations with 10,000 simulation repetitions recommended for each variable (*RM:* 106). The contrast with traditional forms of insurance is marked, as risk modelling for index insurance, as developed by *RM*, is not founded on historical variation. Instead, simulation constructs a system that is predetermined in terms of its variation (on the difference between the 'archival-statistical' mode of traditional insurance and 'enactment-based' knowledge provided by simulations, see Collier 2008).

The use of probabilistic simulations underscores that the reductive objectification of the soil is a process where financial theory is constitutive of the categories used in mapping ecological uncertainties. Here, it is noticeably difficult to separate empirical data from the theoretical models that condition how data is instrumentalised into a tool of weather-related risk prediction (e.g. Edwards 2010, p. 282). The role of the submodels is heavily reduced in the final product. Instead of taking into account environmental factors, it focuses on modelling payout ratios:

> [T]he model is not actually simulating the weather (such as rainfall), nor is it simulating the weather index (for example, drawing from a distribution of index trigger values). Instead, the model directly simulates the uncertainty around the actual payout amounts. An important advantage of this approach is its simplicity and the relative ease of explaining and understanding its results.
>
> (*RM:* 269)

The choice is elaborated by detailing the assumptions and conditions behind successful modelling practices. The suggested strategy presupposes that index insurance operates in isolation from other financial products and, as explained above, only one-year time frames are considered for the payout models. The aim is to predict payout ratios accurately, and this ultimately constitutes the socioeconomic logic of the product. By transforming environmental data into a standardised assemblage of relations, the models construct a form of risk that is approximated by simulating the model's various dimensions as stochastic elements (*RM*: 154–55). In other words, 'risk' here can be claimed to be 'abstract,' as it has nothing to do with, say, the concrete loss of the harvest; rather, it concerns an abstract value derived from the model.

Even if *RM* stipulates payout ratio simulation as the most suitable method for designing index insurance products, it does consider two other approaches for modelling environmental risks, perhaps for didactic reasons. The first is to model the index so that environmental data, such as rainfall, not only serves as a submodel for payout ratio simulations but is also used to simulate dynamic changes in the environment. This approach would make it possible to model sequential relationships between different years and areas and therefore include weather-related changes in the process (*RM:* 270). The second alternative is to model the weather itself. This approach would require a more holistic weather system model, where the product's trigger levels would be formulated on the basis of simulated hazard data instead of historical hazard data. This signifies a much more comprehensive alternative where even the inclusion of multi-year weather cycles, such as El Niño, could be used in the design of an index insurance product (*RM:* 271). Considering these alternative approaches, the authors of the guide weigh better understanding of the world's complexity in relation to instrumental needs:

> In many cases, analysts start off thinking that they need very 'realistic' models to capture the behavior of the real world. However, in our experience it is best to start with the simplest model that fulfills all the needed functions and uses valid assumptions. Only then should analysts add more complexity as necessity dictates.
>
> (*RM*: 271–22)

Hence, the alternative methods are not recommended for index insurance product development. With this, the guide draws a conclusion for the mapping process, basing its ecological risk modelling recommendations on financial performance. The perception of the soil and the weather as static systems is deemed essential for achieving precision and coherence in the pricing of risks. Thus, index insurance, as advanced by *RM*, disregards both real property damages and environmental changes in its technical definition of risks. The most important consequence of this move is that climate change is pushed outside of the range of objects that the models can recognise.

In transforming the soil into an object of governance, index insurance is treating the policy's underlying environmental uncertainties as analogous to market fluctuations. Our analysis contends that, as a technology for considering environmental risks, index insurance follows a logic in which the objectivity of risk modelling is grounded in the instrument's capability to establish formal conditions for market operations. For these purposes, the operationalisation of environmental data plays only a minor role in orienting the model's anticipation of future uncertainties; weather phenomena are simply treated as predetermined variables in probabilistic simulations. Thus, environmental data ends up being operationally more important for transferring risks to the global financial markets than it is for gaining a dynamic view of ecological reality.

The choice to model payout ratios but not the environment or temporal change is presented by *RM* as a necessary control mechanism for approximating short-term risks. The resulting index is a form of information that enables financial services to operate by creating expectations about the future. In mediating economic processes, index insurance is an infrastructure that makes risk taking possible because it allows creditors to price the risks of capital; simultaneously, it creates a distribution channel for financial services. These financial services, for their part, are able to price the risk of capital and thus support the expansion of financial markets.

What makes the form of index insurance advanced by *RM* problematic is that the instrument's models are presented as objective representations of ecological risks, while the mathematical language of probabilistic simulations obscures the process through which the risks are constructed and shaped into social relations. That financial instruments do not merely describe the world but also generate social organisations (LiPuma 2017) is related to the constitution of objects of governance being contingent on infrastructural, political, and cultural configurations (Easterling 2016). The normative design *RM* presents for index insurance development has the potential downside of eliminating the forms of information that would recognise interdependences between social and ecological processes in how risks are shaped. While not recommending it, the guide does raise the question of whether finance-based governance should include environmental data as a factor that structurally orients the model's anticipation of the future. Such modelling techniques might aid understanding how risk technologies are not only managing the soil's risks but also shaping them. This is a point of view that climate change makes all the more important, given the feedback loops between economic processes and ecological systems (e.g. Goodman & Anderson 2020; Moore 2015).

To sum up, it is simply astounding that the index insurance programme does not use environmental data either to predict dynamic changes or to consider underlying uncertainties; this is especially surprising as the programme is highlighted as technologically innovative in the discourse of GIIF, the project out of which *RM* arose. In this regard, the methods used for abstracting weather-related risks from their material reality question the ability of index insurance to respond accurately to climate change. Behind the technocratic hopes for governing weather-related risks, attending to the financial formalisation of environmental data reveals a relatively traditional insurance product.

Conclusions: Bypassing climate change with index insurance

Starting with ecological data and finishing with market analysis, index insurance constructs a form of market transaction that seeks to condense and configure into risks the uncertain relationship that farmers have with the soil. *RM* creates a perspective on how World Bank economists reflect the use of digital infrastructures in deliberate forms of socioeconomic planning. Yet, the reported failures of the index insurance programme also point to difficulties in formalising climate change as risks that can be combined with a functional financial instrument (Angeli Aguiton 2020; Johnson 2021). Other scholars have recently discussed situations where taking into account the dynamics of climate risks is made possible by new and updated models but where the political will to use such 'realistic' models is lacking and conflicts ensue (Elliott 2021; Gray 2021). At the core of political tensions is the question concerning if and how risks that change can be reliably calculated, and if yes, what practical effects it will have that they are taken into account. In this context it is noteworthy that even major reinsurers, such as Munich Re, have recently advertised their capacity to handle expertly 'risks that change' (Lehtonen 2017: 40). This provides an interesting contrast to *RM*, in which the modellers end up suggesting that the dynamics of (climate) change be completely bypassed. A palpable tension ensues. While *RM* presents index insurance as a novel and progressive tool with which environmental hazards can be managed, and while it relies heavily on simulations for providing its knowledge base, that is, the reality that it models is thoroughly 'enacted,' at the same time, its form of dealing with temporality comes close to what Collier (2008) has termed 'archival-statistical' knowledge, characteristic of a traditional form of insurance where it is not taken into account that risks can change.

How index insurance, in the form promoted by *RM*, produces predictions of environmental risks is difficult to justify in terms of climate change-related uncertainties that are already apparent in many places of the world. In *RM*, the soil is made visible and manageable by relying heavily on probability simulations that ignore local relations, even though it is these relations that define how humans depend on their ecological surroundings. Indeed, index insurance in the form advanced by *RM* appears to offer a closed system, even though openness to the changing dynamics of the climate should be underscored.

References

Angeli Aguiton, S 2019, 'Fragile transfers. Index insurance and the circuits of climate risk in Senegal', *Nature and Culture*, vol. 14, no. 3, pp. 282–298.

Angeli Aguiton, S 2020, 'A market infrastructure for environmental intangibles: The materiality and challenges to index insurance for agriculture in Senegal', *Journal of Cultural Economy*, vol. 14, no. 5, pp. 580–595.

Bergson, H, 1896, *Matière et mémoire*, Paris: Félix Alcan.

Bridge, G, Bulkeley, H, Langley, P & van Veelen, B 2020, 'Pluralizing and problematizing carbon finance', *Progress in Human Geography*, vol. 44, no. 4, pp. 724–742.

Chiapello, Eve 2020, 'Stalemate for the financialization of climate policy', *Economic Sociology—The European Electronic Newsletter*, vol. 22, no. 1, pp. 20–29.

Christophers, B, Bigger, P & Johnson, L 2020, 'Stretching scales? Risk and sociality in climate finance', *Environment and Planning A: Economy and Space*, vol. 51, no. 1, pp. 88–110.

Collier, S.J 2008, 'Enacting catastrophe: Preparedness, insurance, budgetary rationalization', *Economy and Society*, vol. 37, no. 2, pp. 224–250.

Collier, S.J & Cox, S 2021, 'Governing urban resilience: Insurance and the problematization of climate change', *Economy and Society*, vol. 50, no. 2, pp. 175–196.

Collier, S.J, Elliott, R & Lehtonen, T-K 2021, 'Introduction: Climate change and insurance', *Economy and Society*, vol. 50, no. 2, pp. 158–172.

Easterling, K 2016, *Extrastatecraft. The Power of Infrastructure Space*, London/New York: Verso.

Edwards, Paul N 2010, *A Vast Machine: Computer Models, Climate Data, and the Politics of Global Warming*, Cambridge, MA: MIT Press.

Elliott, R 2021, 'Insurance and the temporality of climate ethics: Accounting for climate change in US flood insurance', *Economy and Society*, vol. 50, no. 2, pp. 173–195.

Goodman, J & Anderson, J 2020, 'From climate change to economic change? Reflections on 'feedback'', *Globalizations*, vol. 18, no. 7, pp. 1259–1270.

Gray, I 2021, 'Hazardous simulations: Pricing climate risk in US coastal insurance markets', *Economy and Society*, vol. 50, no. 2, pp. 196–223.

Grove, K 2021, 'Insurantialization and the moral economy of ex ante risk management in the Caribbean', *Economy and Society*, vol. 50, no. 2, pp. 224–247.

Jarzabkowski, P, Bednarek, R & Spee, P 2015, *Making a Market for Acts of God. The Practice of Trading in the Global Reinsurance Industry*, Oxford: Oxford University Press.

Johnson, L 2013, 'Index insurance and the articulation of risk-bearing subjects', *Environment and Planning A*, vol. 45, no. 11, pp. 2663–2681.

Johnson, L 2021, 'Rescaling index insurance for climate and development in Africa', *Economy and Society*, vol. 50, no. 2, pp. 248–274.

Johnson, L, Wandera, B, Jensen, N & Banerjee, R 2019, 'Competing expectations in an Index-Based Livestock Insurance Project', *The Journal of Development Studies*, vol. 55, no. 6, pp. 1221–123.

Langley, Paul 2020, 'Assets and assetization in financialised capitalism', *Review of International Political Economy*, vol. 28, no. 2, pp. 382–393.

Lehtonen, T-K 2017, 'Objectifying climate change. Weather-related catastrophes as risks and opportunities for reinsurance', *Political Theory*, vol. 45, no. 1, pp. 32–51.

LiPuma, E 2017, *The Social Life of Financial Derivatives. Markets, Risk, Time*, Durham and London: Duke University Press.

Lucas, C.H & Booth, K.I 2020, 'Privatizing climate adaptation: How insurance weakens solidaristic and collective disaster recovery', *WIREs Climate Change*, vol. 11, no. 6, pp. 1–14.

Mapfumo, S, Groenendaal, H & Dugger, C 2017, *Risk Modeling for Appraising Named Peril Index Insurance Products: A Guide for Practitioners*, Washington, DC: World Bank Publications.

Moore, J.W 2015, *Capitalism in the Web of Life. Ecology and the Accumulation of Capital*, London/New York: Verso.

Ouma, S, Johnson, L & Bigger, P 2018, 'Rethinking the financialization of 'nature'', *Environment and Planning A: Economy and Space*, vol. 50, no. 3, pp. 500–511.

Scott, J.C 1998, *Seeing Like a State. How Certain Schemes to Improve the Human Condition Have Failed*, New Haven: Yale UP.

Taylor, Z.J 2020, 'The real estate risk fix: Residential insurance-linked securitization in the Florida metropolis', *Environment and Planning A: Economy and Space*, vol. 52, no. 6, pp. 1131–1149.

4 Renaturalising sovereignty

Ex-ante risk management in the Anthropocene

Kevin Grove

Introduction

In 1992, Ulrich Beck declared that one defining factor of risk society was the non-insurability of risks. For Beck, the emergence of catastrophic risks such as climate change impacts violated principles of calculability and transferability that had allowed insurers to commodify, price, and transfer risks. Beck's risk society thesis implicitly positioned excess as a limit of insurability: excessive earthly and technological forces could cause catastrophes on previously unimaginable scales. In temporal terms, catastrophes exceed the logic of actuarial calculation, given how they are unprecedented events that cannot be predicted based on past occurrences. In spatial terms, catastrophes affect an entire population, an excessive impact that prevents insurers from spreading risk throughout the population. From this perspective, the relation between insurance and earth may appear straightforward: the earth sets a 'natural' limit on insurance, because insurance cannot provide protection against potentially catastrophic earth system dynamics that are becoming a more common occurrence in the Anthropocene.

However, while this earthly excess may indeed violate traditional forms of actuarial and indemnity insurance, subsequent developments in risk management, insurance and reinsurance, and capital markets have demonstrated the flexibility of insurance as an apparatus for governing uncertain futures (Ewald 1991). Techniques, such as risk pooling, catastrophe modelling, parametric insurance, and weather derivatives, to name but a few, have created new mechanisms for pricing and transferring risk that are based on speculative and enacted forms of knowledge, rather than actuarial and predictive knowledge. Over the past decade, critical insurance scholars have detailed these new insurantial techniques and their biopolitical effects, while drawing attention to the ways they invert key assumptions in Beck's argument. Critical analyses of weather derivatives, for example, demonstrate how these financial instruments generate value out of, rather than in spite of turbulence (Cooper 2010; Martin 2007). Catastrophe models rely on fine-toothed simulations of future catastrophic hurricane impacts to generate probability curves for loss events (Collier 2008). Paired with parametric

DOI: 10.4324/9781003157571-5

insurance contracts, they enable insurance actors to create catastrophe insurance products that leverage planetary volatility into new fields of capital accumulation and state security practice (Lobo-Guerrero 2011; Grove 2012; Johnson 2013).

In this chapter, I will examine the political effects of ex-ante risk management (EARM), a relatively novel form of insurance that has become increasingly influential in development and disaster management fields. In general terms, and as I will detail below, EARM directly refers to a suite of risk management and budgeting tools and techniques, which include various risk transfer instruments such as catastrophe insurance, weather derivatives, contingency funds, and the like. However, EARM also signals a distinct governmental rationality, a way of understanding disaster events as contingent financial liabilities, that seeks to transform how developing states plan for unpredictable financial impacts of catastrophes (Grove 2021). EARM attempts to enable states to 'thin[k] and prepar[e] like an insurance company' (Clarke & Dercon 2016, p. 101). Thinking like an insurer involves approaching disasters as events that generate contingent financial liabilities that developing states need to build financial capacity to manage independently, without the assistance of foreign aid or development lending.

In recent years, development economics have heralded EARM as a paradigm shift in disaster management and development, away from reliance on *ex-post*, relief and response activities, and towards proactive, risk mitigating behaviours (Ghesquiere & Mahul 2007; Wilkinson 2012). To draw out the political effects of this shift, I explore how EARM is recalibrating the capacities and competencies attributed to states and donor agencies in a volatile environment through a case study of Dominican disaster budgeting. In Dominica, the government of Prime Minister Roosevelt Skerrit (the Government of the Commonwealth of Dominica, or GCD) has utilised EARM techniques, such as catastrophe insurance, since the mid-2000s. However, the GCD's efforts to plan for catastrophic fiscal impacts cannot be divorced from the repetition of plantation violence – what Sharpe (2016) calls the plantation's 'wake.' Caribbeanist scholars have demonstrated how vulnerability in the Caribbean involves unresolvable tensions between two parallel but interrelated dynamics: on the one hand, mechanisms and techniques of racialised exploitation and colonial extraction that fuelled modernisation in (formerly) colonial metropoles *through* the production of underdevelopment in peripheral regions; and on the other hand, the ongoing struggle of marginalised peoples – including, at certain times and situations, local elites – to advance alternative modernisation projects (Baptiste & Rhiney 2016; Werner 2016; Thomas 2019).

Development in the Caribbean is thus a tragic site where the modernist promise of sovereignty, autonomy, and independence from colonial control repeatedly runs up against the persistence of the plantation economy that uses mechanisms of debt and structural adjustment to foreclose alternative development pathways and lock in colonial-era dependencies (Scott 2014;

Pugh 2017; Lewis 2020). EARM, as I demonstrate, intensifies these tragic dynamics: while the GCD has deployed EARM as part of its efforts to create a space for political autonomy through strategic budgeting (Grove 2021), donor agency pressures to recalibrate disaster financing around the insurantial logic of EARM is subtly transforming the relation between insurance, the developing state, donors, the public, and a volatile earth in the Anthropocene. The turn to EARM and its imperative to 'prepare like an insurer' leverages potential catastrophic impacts and further hollows out developing states' sovereignty in the Anthropocene.

Defining EARM

Over the past decade, researchers have moved from cautious hints that insurance might promote more equitable adaptation pathways towards full-throated pronouncements that insurance can drive pro-poor adaptation strategies on multiple levels. For proponents, developing states can begin to 'prepare like an insurer' (Clarke & Dercon 2016), combining novel insurance products such as sovereign catastrophe insurance with other EARM tools such as contingency funds or weather derivatives to strategically plan for disaster-induced financial disruptions. 'Thinking like an insurance company' involves apprehending disasters as events that generate *contingent liabilities*, expenditures a state will have to make in the aftermath of a disaster. For example, in many small island developing states (SIDS), disasters often increase demands for welfare services, which draws on resources from states' social protection schemes (World Bank 2017). In effect, this mathematises the state's post-disaster response, recovery, and reconstruction activities, subjecting them to a calculative rationality focused on developing the kind of financial self-sufficiency insurance companies must exercise (Ewald 1991).

To plan for contingent liabilities, insurers rely on a mix of risk retention and risk transfer tools designed to build their financial capacity. Most insurers typically retain a limited amount of risk financed through their capital base, and purchase reinsurance to transfer risk to reinsurance markets. This provides insurers with the capacity to meet their contingent liabilities and avoid bankruptcy, even in extreme loss events. Clarke and Dercon (2016, p. 81) extend this strategy to governments: 'these principles should form the basis of a financial strategy for a government or an organisation committed to covering particular contingent liabilities.' Thus, financing disaster response becomes a matter of financing contingent liabilities without relying on external aid: a government 'must decide how much risk it will retain and how much risk it will transfer, and which financial and budgetary instruments to use for this' (Clarke & Dercon 2016, p. 8). In this view, strategic budgeting can allow developing states to rationalise disaster financing, and thus become financially self-sufficient and capable of managing the financial impact of disasters without relying on external relief (Linnerooth-Bayer & Hochrainer-Stigler 2015; Surminski et al. 2016).

At stake in this insurantialised governmental rationality is the historically specific question of how developing states should prepare for disasters and secure provisioning of vital public services in the face of uncertain and increasingly catastrophic climate change impacts (Johnson 2021). Just as an insurer must prepare for a variety of contingent liabilities that fluctuate wildly from year to year, so too must the state approach its vital infrastructural and welfare services as contingent liabilities with unpredictable costs. And just as an insurer must self-finance these commitments in a way that avoids bankruptcy, so too must the state devise strategies for self-financing its contingent liabilities. Development economists stress that this is particularly important for SIDS, which lack the resources and capacities to raise ex-post funds through channels available to wealthier states, such as tax increases, budget restructuring, or t-bill offers (Ghesquiere & Mahul 2007). SIDS, like insurers, must instead manage the costs of contingent liabilities without ability to generate new sources of income *ex nihilo*.

While the demand for developing states to prepare like an insurer has become quite pronounced in recent years, remarkably, it was largely absent from development economics and disaster management debates until the late 2000s. The next section examines how this paradigm shift was driven by development economists' critical engagements with disaster financing and planetary volatility.

Contextualising EARM

Since the 1970s, the prevailing rationality guiding public investments on proactive risk reduction measures has explicitly *inhibited* pre-event financial planning and investments in EARM technologies. This approach, derived from the influential Arrow-Lind theorem (Arrow & Lind 1970), states that governments should 'ignore uncertainty in public investments and behave as if they were risk-indifferent' – or in other words, that states should avoid investing in EARM as long as their costs exceed their expected benefits (Cummins & Mahul 2009, p. 162). Like all institutional analysis, Arrow-Lind is based on a methodological individualism that assumes 'the social' is an emergent agglomeration of individual choice preferences rather than an ontologically prior object of governmental thought and intervention (Collier 2017). This means that any decisions on provisioning public goods should reflect the sum of individual preferences. Arrow-Lind applies these assumptions to the provision of proactive risk reduction. The theorem asserts that, 'there is a cost of risk-bearing that must be subtracted from the expected return in order to compute the value of the [public] investment to the individual taxpayer' (Arrow & Lind 1970, p. 371). However, when the individual is placed in a large-n population of other taxpayers, the cost of risk-bearing approaches zero, since the amount of the asset that individual claims also approaches zero. The individual effectively becomes risk-indifferent – and the state, as the aggregate of individual preferences, should likewise be indifferent to social risk, and avoid inefficient risk reduction investments.

In the mid-2000s, some development economists began questioning the applicability of these arguments for developing states. Reflecting on Hurricane Ivan's (2004) catastrophic impact on Grenada's economy, they critiqued two assumptions underpinning Arrow-Lind. First, Arrow-Lind assumes that the social cost of risk equals zero if individual losses are not correlated. However, in Grenada, individual losses were highly correlated: the catastrophic losses, totalling nearly 200 per cent of the country's GDP, affected nearly every sector and segment of society, and threatened the state's ability to provide basic services (Grove 2012). Thus, 'small states exposed to natural disasters that can affect the entire country, like small Caribbean islands exposed to hurricanes, face a high social cost of catastrophic risk-bearing' (Ghesquiere & Mahul 2007, p. 7). Second, Arrow-Lind assumes a negligible opportunity cost of post-disaster financing: in theory, states can raise taxes or issue debt rather than shifting expenditures. However, Grenada's excessive losses created immediate liquidity problems that quickly exceeded the state's fiscal capacities. Unable to finance reconstruction through budget reallocations, the state's post-disaster fiscal outlook remained bleak: despite its efforts to generate capacity through donor assistance, debt restructuring, and tax increases, 'Grenada's fiscal situation remained challenging and the country still faced a financing gap of 4.5 percent of GDP for 2005 with total debt projected to increase to 150 percent of GDP' (Ghesquiere & Mahul 2007, p. 17). Thus, 'most developing countries, and particularly small countries, do not have this flexibility in their budget, making the opportunity cost of reserve high, and thus the social cost of catastrophic risk bearing is high' (Ghesquiere & Mahul 2007, p. 8).

Thus, for some development economists, Ivan's impact on the Grenadian state demonstrated significant costs to ex-post relief that Arrow-Lind could not account for. Indeed, just a few years before Ivan, the 2001 Third Assessment Report (TAR) of the Intergovernmental Panel on Climate Change (IPCC) dedicated a chapter to the impacts of climate change on 'insurance and other financial services.' The authors emphasised that the effects of climate change would be greatest in developing countries, cautioning that, 'several countries experience impact on GDP as a consequence of natural disasters; damages have been as high as half of GDP in one case' (Vellinga et al. 2001, p. 420). Ivan surpassed these damages in spectacular fashion, a feat that has been repeated, with devastating consequences, in subsequent years. These events, and their catastrophic impacts, demonstrate a common refrain among critical scholars of the Anthropocene: the *asymmetry* and planetary *indifference* that structures human-earth relations in the Anthropocene (Clark 2010). These arguments emphasise how sublime earthly forces far exceed human capacities to know and control the earth. Rather than a stable backdrop of human activity, the Anthropocene signals how the earth has become both an actor in and the stake of global politics that shapes subjective possibilities and human capacities (Dalby 2020). For some scholars, this new experience of earthly matter, and new experience as earth-bound subjects, creates possibilities for more progressive and benevolent politics (Last 2015; Hawkins 2015; Haraway 2016). For others, the immersion in planetary excess is driving ongoing governance transformations. For example, Chandler (2018) examines how techniques of mapping, sensing, and hacking are reshaping governance practices in ways that make governance responsive and adaptive to earth dynamics – a stark contrast to modernist efforts to impose a human rationality on social and environmental conditions. Lorimer (2020) similarly examines how techniques of probiotic governance utilise excessive powers of life to foster desirable forms of life. These and other governance transformations are not abandoning the modernist desire to control social and environmental outcomes and bend environmental forces to human will, but are rather recalibrating the logics, strategies, techniques, and practices of governance in ways that govern *through* unruly and excessive planetary dynamics (Neyrat 2019; Chandler et al. 2020).

These ongoing transformations of governance in the Anthropocene form the backdrop against which EARM has emerged as a paradigm shift in development and disaster management. Key here is the way development economists drew on new institutional economics (NIE) to understand and critique the dominant Arrow-Lind paradigm in the context of mounting catastrophic losses among Caribbean states and other SIDS. Rather than assuming, as Arrow-Lind does, that states had de facto financial capacity to socialise risk and finance recovery, these economists drew on NIE to assert that 'institutional characteristics and policy choices may have an impact on the macroeconomic consequences of disaster' (Noy 2009, p. 222). As has been well rehearsed, NIE offers a sympathetic critique of neoliberal economic theory (Best 2014; Collier 2017; Grove 2018) that maintains liberal assumptions of methodological individualism, but situates the individual in complex 'environments' that exceed total comprehension. Rationality is thus bounded (Simon 1955) and contingent on the quality of institutions – formal and informal rules, norms and organisational forms that structure decision-making processes (Ostrom 1990) – that mediate between interior worlds of emergent subjective preferences and exterior complex environments that continually present the individual with new information. NIE thus does not assume that a single institution, such as the market, can offer a universally valid measure for determining how to provision public goods and services (Chandler 2014). Instead, decision-making in conditions of imperfect knowledge involves an indeterminate and reflexive process of institutional design, a creative and synthetic act of 'tinkering' and experimenting with different institutional forms to assemble competing interests into contingent, pragmatic solutions to intractable problems that lack a single, optimal solution (Ostrom 1999).

The World Bank's country catastrophe risk management framework exemplifies this rationality. The framework assists developing state governments' efforts to design financial planning strategies appropriate to their specific risk financing needs and risk profiles (Ghesquiere & Mahul 2010). In brief, it differentiates disaster risks into three layers, ranging from low to high impact and high to low frequency, respectively, and presents these in relation to a 'menu of financial instruments and mechanisms' that 'governments can choose from' (World Bank 2017, p. 20). Contingency funds or budget reallocations can finance low-impact, high-frequency events such as localised floods. Concessional lending or contingent debt instruments can finance medium-impact, medium-frequency events such as larger floods or small earthquakes. Catastrophe insurance or catastrophe bonds can finance high-impact, low-frequency events such as major hurricanes (Ghesquiere & Mahul 2010). Each of these EARM instruments has specific functional characteristics: budget reallocation, contingency funds and some parametric insurance products tend to be highly liquid, and are thus appropriate for meeting immediate relief needs in the first three months following a catastrophic event. Contingent debt instruments and catastrophe bonds have moderate liquidity, which makes them more appropriate for recovery and reconstruction activities (three to nine months). In contrast, ex post financing instruments, such as domestic and external credit, donor assistance, foreign aid, or tax increases, generally have low liquidity and thus typically cannot be accessed for six to nine months. Relying on ex post mechanisms can slow disaster relief and reconstruction, compounding a disaster's impacts.

'Thinking like an insurer' thus involves reimagining SIDS' budgeting practices in the terms and techniques insurers utilise to become financially self-sufficient. At first glance, this may intensify a neoliberal push to financialise disaster management (Grove 2012). However, Collier's examinations of neoliberal budgetary reforms in post-Soviet Russia caution against blanket categorisations. Drawing on Rose's (1999) emphasis on the formal character of neoliberalism, Collier (2005, p. 375) shows how neoliberal budgetary reforms involve an indeterminate process of formal rationalisation, or 'increasing... the extent of quantitative calculation that is technically possible and actually exercised in determining the allocation of resources in a given society or social system.' Collier's analysis raises the question of *how* specific budgeting and financial planning strategies manage the tensions, juxtapositions, and contradictions between the formal rationalisation of the state's substantive goals, and the existing institutional and infrastructural arrangements that structure the particular forms of life state biopolitics takes as its object. This is particularly salient for Dominican disaster budgeting, where the GCD's adoption of EARM techniques reflects ongoing transformations in sovereignty driven by both the problem of financial capacity in the Anthropocene and the illusory pursuit of autonomy in the post-independent Caribbean. The next section examines how EARM is subtly transforming Dominican sovereignty in the Anthropocene.

Renaturalising sovereignty in the Anthropocene

In Dominica, despite promises of autonomy and freedom attached to political independence, the GCD has long grappled with the challenge of fulfilling its biopolitical mandate while meeting externally imposed fiscal management measures. Prior to its 1978 independence, anti-colonial and nationalist activists argued that political independence was the essential precondition for economic self-sufficiency. These arguments, variants of Marxist dependency theory and pragmatic economic nationalism, suggested that independence would allow the state to engage international markets on its own terms, rather than those of the British colonial office (André & Christian 2002). Set in their context, there was much to this argument: British colonial administrators actively inhibited industrialisation across the Caribbean in order to protect consumer markets for British manufacturers. Dominica's preferential trading status within European banana markets also created an illusory sense of economic self-sufficiency that made political independence economically viable (Baker 1994). However, the WTO's 1999 'banana wars' ruling shattered this illusion: Payne (2008, p. 300) emphasises that, 'after all, Dominica's crisis came to a head following the decision of a WTO dispute panel at which the Caribbean banana-producing countries were not even permitted to be present.' Events such as the WTO ruling, and global recessions in 2001–2003 and 2008 repeatedly drove home the hollowness of political sovereignty in the contemporary global political economy (Bonilla 2015). Recent disaster events have compounded these fiscal struggles: while 2015's Tropical Storm Erika caused economic losses in excess of 50% of the state's GDP, the back-to-back category 5 Hurricanes Irma and Maria (2017) caused damages of 224% of the island's GDP (Grove 2021).

Faced with negative growth and expanding debt, in 2002 the GCD entered into a Structural Adjustment Programme (SAP) with the IMF. The SAP required macroeconomic stabilisation and long-term adjustment policies designed to reduce fiscal imbalances and orient the economy towards long-term growth. It also mandated that the government prepare regular Poverty Reduction Strategies Papers (PRSP), a common IMF and WB technique to build local ownership of growth-oriented poverty reduction development initiatives (Best 2014). The GCD's first PRSP was prepared in 2004; subsequently, it rolled the PRSP into its master economic development plan, the Growth and Social Protection Strategy (GSPS), published in 2006, 2008, 2012, and 2014. The GSPS expresses the 'imperial logic of growth' (Sealey-Huggins 2017) that ties poverty reduction and sustainable development to limitless growth: the inaugural GSPS emphasised that, 'the [GCD] intends to conduct prudent fiscal policy that is conducive to growth, based on expenditure restraint, administrative modernizing and reform, and control of borrowing in a manner that reduces government debt to sustainable levels' (GCD 2006, p. 4).

I have detailed wider trends in Dominican budgeting elsewhere (Grove 2021); for now, what matters is that the Dominican government has adopted an array of EARM techniques since its 2002 IMF SAP to manage both its financial exposure to more intense disaster events in the Anthropocene and its exposure to the pressures of neoliberal structural adjustment lending. Consistent with the WB's country catastrophe risk management framework described above, these involve a mix of insurance and risk management techniques. For example, the GCD has been a member of the Caribbean Catastrophic Risk Insurance Facility (CCRIF) since the Facility's 2007 launch. More recently, it also participated in the Disaster Vulnerability Reduction Project (DVRP), financed through the WB's Climate Investment Fund. Along with providing funding for pre-event risk mitigation projects, the DVRP allows up to US$1 million to be diverted from capital projects to finance short- and medium-term response and recovery efforts (World Bank 2018). The GCD also launched a national disaster contingency fund, the Vulnerability, Risk and Resilience Fund (VRRF). The VRRF is form of self-insurance that provides funds targeted to long-term reconstruction and rehabilitation initiatives.

The design of the VRRF demonstrates how EARM can paradoxically erode the possibilities for autonomy in the Anthropocene. Each of the GCD's pre-Erika GSPS documents included plans and promises for the creation of a national disaster contingency fund. Alternatively called the 'National Disaster Contingency Fund,' 'Disaster Mitigation Contingency Fund,' or 'Environmental Mitigation Fund,' the goal was the same: to provide the government with immediate post-disaster liquidity to 'cover the costs of repairs and environmental enhancements' (GCD 2006, p. 84.). Importantly, each proposal recommended capitalising the fund through the Public Service Investment Programme (PSIP). The PSIP is a targeted capital expenditure program focused on developing the island's critical infrastructure systems, especially roads, bridges, utilities, and telecommunications. The PSIP is funded through a mix of grants (65–75 per cent, depending on the year), loans (5–10 per cent), and the government's own funds (20–25 per cent). However, the uncertain amount of foreign aid the GCD receives from year-to-year resists formal rationalisation. In effect, funding the VRRF through the PSIP would have inhibited the GCD's ability to prepare like an insurer, since its financial capacity would always be contingent on donor aid revenue. In contrast, following Tropical Storm Erika, in 2016 IMF advisors recommended capitalising the VRRF through the island's Citizenship By Investment (CBI) programme (Guerson 2016), a significant source of locally generated revenue. The government eventually adopted the IMF's suggestions and capitalised the VRRF through the CBI.

The Fund's recommendation attempts to responsibilise the GCD on two fronts. First, it compels the GCD to use its own internally generated revenues, which are more amenable to formal rationalisation. Second, to further compel the GCD to prepare like an insurer, the IMF's recommendations also included the need for 'unambiguous budget contribution and disbursement rules, with triggers based on verifiable criteria, a clearly-stated objective, and strict information disclosure requirements to ensure the transparency of its operations' (Guerson 2016, p. 13). For development economists, designing-in triggers to disaster response plans provides the kinds of coercive rules new institutionalist economists argue are required to lock-in both donor and recipient decisions in advance of disaster events (Clarke & Dercon 2016, p. 65). Triggers depersonalise and algorithmically structure decision-making processes in advance of a disaster event, and thus remove the subjective and institutional elements of disaster management that can distort decision-making processes, delay rapid disbursements of funds, and prevent the government from managing disaster response in a timely and efficient manner.

Set in the context of struggles for political autonomy in the post-independence Caribbean, the insurantialisation of governance through the formal rationalisation of development and disaster budgeting and the use of triggers produces two significant state effects. First, insurantialisation reconfigures the boundaries between the developing state, donors, markets, and the public, in the process transforming the distribution of capacities, competencies, responsibilities, and expectations that structures how decisions on pre- and post-disaster financing can be made in the face of climate change impacts. Triggers attempt to fully automate the decision-making process, in the name of achieving a more efficient and self-financed disaster response. Clarke and Dercon (2016) suggest that:

> to work as an index insurance product, a trigger should not lead to a set of options for a decision-making body; it should result in an automatic decision. In other words, a defined set of indicators reaching particular pre-agreed values should lead to a defined action, as in insurance.

In effect, insurantialisation relocates the nominally sovereign decision over how to manage the population's welfare from the political realm, the province of the developing state, to the technical sphere (Aradau & van Munster 2011). Decisions on how to utilise limited government revenue to address competing demands for immediate biopolitical needs and anticipatory, pre-event risk management investments, and decisions on when and how to access and distribute post-event contingency funds, becomes an automated effect of pre-determined indicators developed in consultation between the GCD, strategically courageous donors, and IMF advisors that lock in the government's actions in advance of any actual disaster event. This new configuration of developing states, donors, markets, and the public further hollows out the GCD's effective sovereignty, relocating the sovereign decision from the centralised state to a disaggregated risk assemblage of technical devices, contracts, regulations and development consultants that structure triggering mechanisms.

Second, insurantialisation also reconfigures the relation between the state, donors, markets, the public, and the earth. Rather than a stable backdrop on which sovereignty is performatively enacted, the earth itself becomes an active limit on sovereignty: it is what the post-independence Caribbean state must abide by and respond to. EARM effectively renaturalises sovereignty in the Caribbean. Instead of signalling a capacity to transcend the earth, in the Anthropocene, the use of triggers to determine when and how state resources can be accessed and how they can be used immerses the state within planetary dynamics that can disrupt its core biopolitical functions. The insurantialisation of disaster governance is thus a moral response to the challenge of earthly excess in the Anthropocene that intensifies rather than ameliorates colonial-style geopolitical dependencies, for it makes the state's ability to sense and respond to planetary dynamics contingent on its adoption of EARM strategies and techniques. Set in the context of 400 years of plantation violence, the cost of these novel sensory and response capacities is incalculable, for it requires Caribbean states to sacrifice the limited and precarious forms of autonomy won through centuries of anti-colonial struggle.

Conclusions

In the Anthropocene, experiences of excessive earthly powers have been animating transformations in political subjectivity as well governance strategies and tactics. This chapter has explored how the conjunction of asymmetrical planetary excess and insurantial strategies are reconfiguring state-society-economy relations in the Anthropocene. More intense tropical cyclones, and the increasingly excessive disaster losses, threaten SIDS, such as Dominica, with fiscal insolvency. Development economists have seized on these conditions to critique ex-post disaster relief, and instead advocate for EARM measures that enable states to 'prepare like an insurer.' EARM initiatives attempt to make disaster budgeting more reflexive and responsive to emerging planetary dynamics, and thus maintain economic independence and manage the costs of disaster events without relying on foreign aid. However, as the example of Dominican disaster budgeting demonstrates, in practice, EARM renaturalises sovereignty as it attempts to recalibrate how states utilise financial resources to prepare for and respond to emergency and excessive planetary forces. The formalised rationality of EARM prioritises future economic self-sufficiency over present autonomy. In Dominica, EARM initiatives such as the VRRF reconfigure the government's budgeting practices around insurantialised governance strategies that further distance the GCD from decisions on when and how to use its limited government revenues. In effect, the moral imperative to become economically self-sufficient in the face of mounting climate change impacts presents

Caribbean governments with a tragic choice, for building financial capacity by preparing like an insurer requires sacrificing hard-won autonomy.

References

André, I & Christian, G 2002, *In search of Eden: Essays on Dominican history*, Bowie, MD: Ponde Casse Press.

Aradau, C & van Munster, R 2011, *The politics of catastrophe: Genealogies of the unknown*, Abingdon: Routledge.

Arrow, K & Lind, R 1970, 'Uncertainty and the evaluation of public investment decisions', *The American Economic Review*, vol. 60, no. 3, pp. 364–378.

Baker, P 1994, *Centering the periphery: Chaos, order and the ethnohistory of Dominica*, Montreal: McGill-Queen's University Press.

Baptiste, A & Rhiney, K 2016, 'Climate justice in the Caribbean: An introduction', *Geoforum*, vol. 73, no. 3, pp. 17–21.

Best, J 2014, *Governing failure: Provisional expertise and the transformation of global development finance*, Cambridge: Cambridge University Press.

Bonilla, Y 2015, *Non-sovereign futures*, Chicago: University of Chicago Press.

Chandler, D 2014, *Resilience: The art of governing complexity*, Abingdon: Routledge.

Chandler, D 2018, *Ontopolitics in the Anthropocene: Mapping, sensing, hacking*, Abingdon: Routledge.

Chandler, D, Grove, K & Wakefield, S (eds.) 2020, *Resilience and the Anthropocene: Governance and politics at the end of the world*, Abingdon: Routledge.

Clark, N 2010, *Inhuman nature: Sociable life on a dynamic planet*, London: Sage.

Clarke, D & Dercon, S 2016, *Dull disasters? How planning ahead will make a difference*, Oxford: Oxford University Press.

Collier, S 2005, 'Budgets and biopolitics', In S Collier & A Ong (eds.), *Global assemblages: Technology, politics and ethics as anthropological problems*, Oxford: Blackwell, pp. 373–390.

Collier, S 2008, 'Enacting catastrophe: Preparedness, insurance, budgetary rationalization', *Economy and Society*, vol. 37, no. 2, pp. 224–250.

Collier, S 2017, 'Neoliberalism and rule by experts', In V Higgins & W Larner (eds.), *Assembling neoliberalism: Expertise, practices, subjects*, New York: Palgrave, pp. 23–43.

Cooper, M 2010, 'Turbulent worlds: Financial markets and environmental crisis', *Theory, Culture & Society*, vol. 27, no. 2–3, pp. 167–190.

Cummins, J & Mahul, O 2009, *Catastrophe risk financing in developing countries: Principles for public intervention*, Washington, DC: The World Bank.

Dalby, S 2020, *Anthropocene geopolitics: Globalization, security, sustainability*, Ottawa: University of Ottawa Press.

Ewald, F 1991, 'Insurance and risk', In G Burchell, C Gordon & P Miller (eds.), *The Foucault effect: Studies in governmentality*, Chicago: University of Chicago Press.

GCD (Government of the Commonwealth of Dominica) 2006, *Medium-term growth and social protection strategy*, Roseau: GCD.

Ghesquiere, F & Mahul, O 2007, 'Sovereign natural disaster insurance for developing countries: a paradigm shift in catastrophe risk financing', *Policy research working paper #4345*, Washington, DC: The World Bank.

Ghesquiere, F & Mahul, O 2010, 'Financial protection of the state against natural disasters', *World Bank policy research working paper #5429*, Washington, DC: The World Bank.

Grove, K 2012, 'Preempting the next disaster: Catastrophe insurance and the financialization of disaster management', *Security Dialogue*, vol. 43, no. 2, pp. 139–155.

Grove, K 2018, *Resilience*, Abingdon: Routledge.

Grove, K 2021, 'Insurantialization and the moral economy of ex ante risk management in the Caribbean', *Economy & Society*, vol. 50, no. 2, pp. 224–247.

Guerson, A 2016, 'Assessing government self-insurance needs against natural disasters: An application to the ECCU', *IMF Country Report 16/333*, Washington, DC: IMF.

Haraway, D 2016, 'Anthropocene, capitalocene, plantationocene, chthulucene: Making kin', *Environmental Humanities*, vol. 6, no. 1, pp. 159–165.

Hawkins, H 2015, *Creativity*, Abingdon: Routledge.

Johnson, L 2013, 'Catastrophe bonds and financial risk: Securing capital and rule through contingency', *Geoforum*, vol. 45, pp. 30–40.

Johnson, L 2021, 'Rescaling index insurance for climate and development in Africa', *Economy and Society*, vol. 50, no. 2, pp. 248–274.

Last, A 2015, 'Fruit of the cyclone: Undoing geopolitics through geopoetics', *Geoforum*, vol. 64, pp. 56–64.

Lewis, J 2020, *Scammer's yard: The crime of black repair in Jamaica*, Minneapolis: University of Minnesota Press.

Linnerooth-Bayer, J & Hochrainer-Stigler, S 2015, 'Financial instruments for disaster risk management and climate change adaptation', *Climatic Change*, vol. 133, pp. 85–100.

Lobo-Guerrero, L 2011, *Insuring security: Biopolitics, security and risk*. Abingdon: Routledge.

Lorimer, J 2020, *The probiotic planet: Using life to manage life*, Minneapolis: University of Minnesota Press.

Martin, R 2007, *An empire of indifference: American war and the financial logic of risk management*, Durham: Duke University Press.

Neyrat, F 2019, *The unconstructable earth: An ecology of separation*, New York: Fordham University Press.

Noy, I 2009, 'The macroeconomic consequences of disasters', *Journal of Development Economics*, vol. 88, pp. 221–231.

Ostrom, E 1990, *Governing the commons: The evolution of institutions for collective action*, Cambridge: Cambridge University Press.

Ostrom, E 1999, 'Coping with the tragedies of the commons', *Annual Review of Political Science*, vol. 2, pp. 493–535.

Payne, A 2008, 'After bananas: The IMF and the politics of stabilisation and diversification in Dominica', *Bulletin of Latin American Studies*, vol. 27, no. 3, pp. 317–322.

Pugh, J 2017, 'Postcolonial development, (non)sovereignty and affect: Living on in the wake of Caribbean political independence', *Antipode*, vol. 49, no. 4, pp. 867–882.

Rose, N 1999, *Powers of freedom: Reframing political thought*, Cambridge: Cambridge University Press.

Scott, D 2014, *Omens of adversity: Tragedy, time, memory, justice*. Durham: Duke University Press.

Sealey-Huggins, L 2017, '"1.5°C to stay alive": Climate change, imperialism and justice for the Caribbean', *Third World Quarterly*, no. 38, pp. 2444–2463.

Sharpe, C 2016, *In the wake: On blackness and being*, Durham: Duke University Press.

Simon, H 1955, 'A behavioral model of rational choice', *The Quarterly Journal of Economics*, vol. 69, no. 1, pp. 99–118.

Surminski, S, Bouwer, L & Linnerooth-Bayer, J 2016, 'How insurance can support climate resilience', *Nature Climate Change*, vol. 6, pp. 333–334.

Thomas, D 2019, *Political life in the wake of the plantation: Sovereignty, witnessing, repair*, Durham: Duke University Press.

Vellinga, P, Mills, E, Berz, G et al. 2001, 'Insurance and other financial services', Working Group 2, Chapter 8, In *Climate change 2001: Impacts, vulnerability, and adaptation*, Geneva: IPCC.

Werner, M 2016, *Global displacements: The making of uneven development in the Caribbean*, Oxford: Wiley-Blackwell.

Wilkinson, E 2012, *Transforming disaster risk management: A political economy approach*, ODI Background Note, London: Overseas Development Institute, January 2012.

World Bank 2017, *Sovereign climate and disaster risk pooling*, World Bank Technical Contribution to the G20, Washington, DC: The World Bank.

World Bank 2018, *Dominica – disaster vulnerability reduction project: Additional financing*, Washington, DC: The World Bank.

Section II

Water

Thales, considered the first philosopher in Ancient Greece, identified water as *the* basic element on Earth and in the universe. In most religions, water embodies purity and performs an important role in purification, though its transmogrifications and malleability render it far from unilateral, far from absolutely catharising. Being at the mercy of water, encompassed in it, tossed by it, or drowned beneath it, evokes a primal fear.

Existing as liquid, gas, and solid, water's forms and capacities are defined through its constitution with other elements and processes. Air-borne condensed water may refract sunlight producing rainbows, when high density of water vapour meets a cool surface dew forms, suspended and mobile droplets cluster as clouds, and falling in a flurry, intricate, and divisible flakes congregate as layers, drifts, and plumes of snow.

The cycling of water through precipitation, evaporation, transpiration, and runoff is essential to life on earth. This cyclical processing has dictated the distribution of settlements large and small, the distribution and maintenance of arable lands, the boundaries and shift of ecosystems, and the creation of landscapes through erosion and deposition.

Domesticated, we dive, fish, surf, and swim in it and boil, simmer, steam, and pressure cook with it. From the tap or bottle, we imbibe it alone, mixed, melting, or carbonated. But unleashed – even on the micro-scale of a dripping household tap – it threatens homelife and liveability and can wreak environmental, financial, and material havoc. Water has instigated what many see as the most costly, and certainly the most researched form of insurance – flood insurance (Lucas et al. 2021).

While lightning is said never to strike the same place twice, once a place is flooded, it will almost inevitably be so again. Relative to other extreme weather events, topography and water flows are calculable and predictable, leading to a quantitative approach in much of the research on flood insurance. The methodologies of the majority of the research literature discursively frame insurance as a risk management tool that is benign in its constitution and useful if properly and rationally applied (Lucas et al. 2021). There is concern for how premium prices and risk perceptions influence uptake rates, and the relationship between insurance uptake and mitigation

DOI: 10.4324/9781003157571-6

decision-making. In this, the capacities and qualities of water are reduced through flood calculations and modelling, rendering water and insurance concrete and constant.

This enacts types of spatiality that appear at odds with the dynamic nature(s) of water. Instead, Law and Mol (2001) describe a fluid spatiality in which things that may appear at first glance static and stable, change shape. As shape-changers, things in their mobile forms can take their place at different sites and in different ways. While still retaining their status as particular things or objects, 'there is a process of gradual adaptation. Shape invariance is secured in a fluid topology in a process of more or less gentle flow' (Law & Mol 2001, p. 614).

The tentative roll-out of micro-insurance to subsistence farmers in the developing world appears malleable or mutable in its framing of insurance (Booth 2021). In encountering local knowledge, this insurance can adapt and change its form and function. Insurance enters as actuarial and morphs in response to traditional knowledge systems. For example, goat entrails used in weather divining form part of the rationale in insurantial decision-making in some parts of Africa (Johnson 2013). Micro-insurance may retain its status as insurance, yet it also flows between farms, villages, and regions because of its capacity to change and adapt.

Empirically, this fluidity can be a source of the success of an object or thing. Theoretically, fluid spatiality 'suggests that varying configurations, rather than representing breakdown and failure, may also help to strengthen objects' (Law & Mol 2001, p. 615). There appears significant scope for considering the on-going evolution and success of insurance in this light: how, for example, insurance responds to the transmogrifying and malleable properties of water – as sustenance or catastrophe, and how the insurance sector may enact marketised experimentation using fluid configurations of insurance products.

In this section, our authors approach insurance in the face of unwanted large quantities of water that surge through towns and cities, and cascade into home and offices. Rebecca Elliot begins the section considering flood insurance and the 'flow' or fluidity of cross-subsidisation between low and high risk households in the United States and the United Kingdom. This sharing of risk costs adapts and changes with technological and normative shifts and, as Elliot observes, is it an essential part of insurance despite free-market expectations of full risk pricing. In relation to insurance of flooding, Chloe Lucas and Travis Young compare household experiences in Australia and the United States, and identify similarities and differences in resilience and vulnerability manifest through two different insurances – one private and one public. In the third chapter of the section, Mark Kammerbauer and Christine Wamsler also draw attention to issues of insurance-related vulnerability, this time in flood-prone Germany. There may be fluidity in cross-subsidisation and household experience, yet insurance mechanisms can appear inadequate in supporting adaptive changes in the face of intensifying flood risk.

Overall, while each type of flood insurance discussed in this section is premised on a similar set of insurance logics, these logics are mobile and adaptable as they flow and circulate in different places and contexts. This adaptive fluidity, as Law and Mol (2001) observe, can act to strengthen things such as insurance. However, this does not mean that vulnerabilities – or schisms – are not also apparent.

References

Booth, K 2021, 'Critical insurance studies: Some geographic directions', *Progress in Human Geography*, vol. 45, no. 5, pp. 1295–1310.

Johnson, L 2013, Index insurance and the articulation of risk-bearing subjects', *Environment and Planning A: Economy and Space*, vol. 45, pp. 2663–2681.

Law, J & Mol, A 2001, 'Situating technoscience: An inquiry into spatialities', *Environment and Planning D: Society and Space*, vol. 19, pp. 609–621.

Lucas, C, Booth, K & Garcia, C 2021, 'Insuring homes against extreme weather events: A systematic review of the research', *Climatic Change*, vol. 165, no. 3, Article 61.

5 Stopping the flow

The aspirational elimination of flood insurance cross-subsidies in the United States and the United Kingdom

Rebecca Elliott

In March 2014, *The Economist* published a short and highly critical piece about the ways flood risk is priced and pooled through insurance in the United States and the United Kingdom. Though the two countries rely on quite different arrangements – a fully public program in the United States and a private market backed up by a state-organised not-for-profit fund in the United Kingdom – both systems were subject to reproach for offering policies that 'subsidise and pool flood risks instead of pricing them in the market' (Anonymous 2014, p. 76). For the anonymous writers at the magazine, this subsidisation was indeed a 'crime,' principally because it was ostensibly blunting incentives that discourage building in flood-prone areas by keeping rates in high-risk areas artificially low. 'For flood insurance,' *The Economist* concluded, 'a problem shared may in fact be a problem doubled—or worse.'

A cross-subsidy exists when one group of policyholders, for one reason or another, is charged higher premiums so that another group will have lower premiums. This departs from strict actuarial rating, where everyone pays according to their risk. *The Economist* is hardly the first or lone voice pointing to seemingly insidious effects of cross-subsidisation and risk pooling in flood insurance. In recent years, particularly as the extant and expected effects of climate change have come into view, an array of researchers, journalists, environmentalists, insurance and reinsurance interests, think tanks, and others have argued for reforms that bring flood insurance arrangements more in line with actuarial orthodoxy – that is, establishing rates that reflect individual risk transfer without cross-subsidisation (Kousky & Shabman 2014).

Yet, risk-sharing and the pooling of resources, of one type or another, define insurance. Insurance is indeed predicated on 'a problem shared,' as *The Economist* put it. What's more, 'some cross-subsidies will be present in any insurance program' (Kousky 2018, p. 25). From the perspective of the policyholder, taking out insurance may seem like engaging in a kind of personal savings, where one's own funds are put away so they are available in the future. But in fact, insurance works by creating forms of 'collective mutuality,' where participants in the insurance scheme agree to contribute

DOI: 10.4324/9781003157571-7

funds that will compensate the losses of any one of its members (Baker & Simon 2002; Lehtonen & Liukko 2015; Stone 2002). What the controversy over existing cross-subsidies suggests, then, is that there is something at stake in the matter of precisely how, on what terms, and with what effects this takes place.

This chapter analyses aspirational projects to minimise or eliminate flood insurance cross-subsidisation in the US and UK. My approach conceives of a cross-subsidy as a kind of imagined 'flow,' where resources seem to move around across insureds, as the revenue collected in one place makes it possible to continue building and living in another. It cognitively and financially connects people who may have different orientations to and experiences of living near the water. Efforts to minimise or eliminate cross-subsidies from insurance arrangements frame this flow as a problematic one: there is too much of something – risk, responsibility, reward – in one place and too little of it in another. This flow needs correction. It needs to be reduced, stopped, or redirected in ways that set normatively desirable boundaries – floodwalls, say – around what people, private firms, or governments put in and what they get out.

Insurance institutions reflect and enact the political cultures and moral economies in which they are embedded. You can tell a lot about what a society values, and how it defines the relationships between citizens, the state, and the market, based on what it insures and how (Baker 1996; Collier 2014; Elliott 2021). In some countries, such as France, Spain, and Denmark, where 'solidarity system[s]' exist, cross-subsidisation is regarded as a social good. That premiums are decoupled from individual risks, that flows exist between insureds, is not regarded as inappropriate, or at least not so inappropriate that it would outweigh the greater good of making sure everyone can access affordable coverage (Lamond & Penning-Rowsell 2014).

In the United States and the United Kingdom, by contrast, to the extent that cross-subsidisation continues in actually existing insurance arrangements right now, many decision-makers and observers consider this to be a temporarily tolerated, and functionally or politically necessary, evil. In both countries, plans for the future of flood insurance seek to minimise or eventually eliminate cross-subsidisation, even as both systems continue to rely on it to achieve certain aims. In the United States, the latest effort to minimise cross-subsidies is framed as a technical achievement that will immediately realise more *equitable* conditions for policyholders. In the United Kingdom, a more gradual transition away from cross-subsidies is framed as a political achievement that will realise more *prudent* outcomes, redistributing responsibilities across individuals, the government, and the private market.

Flooding is already the costliest 'natural' disaster in both countries, and both countries are facing higher expected risks and losses due to further climate change. Looking at the two cases together, I argue, illuminates a core contradiction at the heart of these aspirational projects as they are pursued in the context of climate change: namely, that addressing the flow that matters – the water that goes where it isn't wanted – will require flows of resources and responsibility no matter what. Eliminating cross-subsidisation does not eliminate social interdependence.

The United States: Risk rating 2.0 and the technical achievement of 'equity'

Because most flood insurance in the United States is provided by a public, federal program – the National Flood Insurance Program (NFIP) – changes to rate setting have been politically conspicuous and often controversial. Since 2012, reforms to the NFIP have focused on moving all insureds closer to individual risk-bearing, meaning their premiums should be based on actuarial rating of their assessed exposure to risk. In fact, that is how the program was initially designed and expected to work when it was established by Congress in the 1960s. An actuarial program of flood insurance, its architects believed, would incentivise risk-mitigating action among those already in the floodplain, and disincentivise any further uneconomical use of floodplains that had yet to be developed. However, from the start, some forms of cross-subsidisation have been practically unavoidable, largely in the interest of making flood insurance affordable and therefore accessible and desirable to those who needed it most (Elliott 2021).

The latest and current effort to minimise cross-subsidisation in the NFIP is called 'Risk Rating 2.0,' and it promises to put all properties 'on a glide path to actuarial rates' (Horn 2021, p. 10) where there currently exist various discounts and cross-subsidies. The Federal Emergency Management Agency (FEMA), which administers the NFIP, describes Risk Rating 2.0 as a technical transformation – a 'transformational leap forward' (FEMA 2021a) – that will yield socially beneficial effects. Because FEMA has 'updated' and 'improved' the underlying technology behind assessing risk, cross-subsidisation is in a sense no longer needed. Cross-subsidisation was, rather, an unfortunate by-product of the limitations of FEMA's historic approach to documenting and quantifying flood risk, which can now be made obsolete.

That historic approach calculated premiums based on broad rating classes – specifically, flood zones on FEMA's 'flood insurance rate maps' (FIRMs). You and your neighbour would share a designation as living in a high-risk zone on the FIRM, and your insurance premiums would reflect that shared designation. This involved an implicit cross-subsidy within the zone, to the extent that flood risk varies within it. For example, the people at the edge of a coastal zone closest to the water will presumably face higher flood risk than those located further inland, but that relatively higher risk is not reflected in rating if they share the same broad zone. Under Risk Rating 2.0, the map-based rating, with its broad rating classes, is to be replaced by rating that 'will reflect *each building's individual flood risk* using structure-specific data that are easier to understand' (FEMA 2021b, p. 5, emphasis added). Though any area's flood risk is a collectively and historically produced problem (Koslov 2019), each individual insured is imagined as having dominion over their own portion of it. The new 'structure-specific' data includes the property's distance and elevation relative to a flood source and other characteristics of the building itself, along with more flood risk variables (e.g. flood frequencies and types) (Horn 2021).

This tighter classification of risk leads FEMA to claim that 'Risk Rating 2.0 is equity in action.' What is equity, as it's framed here? Principally, the end of cross-subsidisation will signal the end of 'unfair' flows between different types of policyholders. No longer will the premiums paid by a lower-risk policyholder – someone located further from the water or built at higher elevation – include some measure of additional cost that makes it possible for their higher-risk neighbour to continue to afford their premiums. 'Individuals will no longer pay more than their fair share in flood insurance premiums' (FEMA 2021b, p. 2).

This idea of a 'fair share' reflects and coheres with ideas about 'actuarial fairness': a term of art in insurance that 'foregrounds individuals and automates accountability' (Kiviat 2019, p. 1138; see also Landes 2015). Higher-risk insureds pay higher premiums; this is their 'fair share.' Lower-risk insureds 'deserve' to pay lower rates because they introduce less risk to the pool and are less likely to make claims. As Tom Baker (2002, p. 395) observes, 'classifying insureds according to risk both reflects and creates a moral vision' – in this case, one where it seems essentially fair that people should only bear responsibility for the quantity of risk they are in a sense imagined to 'own.' This is true even in the case of flood risk, where exposure to the risk and one's ability to manage it at the individual level are profoundly determined by decisions and developments that transcend the property owner's control (e.g. the existence and maintenance of structural flood protection, continued real estate development around one's home, etc.) and in many cases predate their decision to move into a floodplain (e.g. zoning that permitted home building, the construction of public infrastructure to serve the property, etc.). Insurance rating based on risk classification helps to 'persuade people that the purpose of insurance is individual protection and, accordingly, that the insurance group is a collection of individuals without any responsibility to one another' (Baker 2002, p. 395).

Ideas about actuarial fairness have persisted throughout the NFIP's history, informing arguments for reform since long before Risk Rating 2.0. But the practical realisation of a fully actuarial footing for the program has been circumscribed by other ideas of fairness that have normatively and politically justified continued cross-subsidisation (Elliott 2017). For instance, a 'grandfathering' provision has worked such that, when a FIRM is updated, if a property is remapped into a higher flood risk rate class, the policyholder can retain their older, lower rate. Even though the long-term effect of grandfathering is that, as maps are updated, increasing numbers of policies violate *actuarial* fairness, this provision was adopted with 'public goals in mind, namely maintaining participation in the program and a perception of fairness on the part of homeowners in response to updated mapping' (Kousky & Shabman 2014, p. 8). The idea was that it was only fair that the program not penalise property owners who had, in a sense, played by the rules, by building to code and maintaining coverage, because the world had changed around them and the maps had been updated to reflect that. The grandfathering discounts are offset by other policyholders in the zone, who pay higher rates (Horn 2021), with the NFIP increasing the cross-subsidy from non-grandfathered policyholders to account for the lost revenue (Committee on the Affordability of National Flood Insurance Program Premiums 2015). For decades, these two ideas of what fairness requires have uneasily co-existed in the NFIP. Risk Rating 2.0 seems to eliminate this longstanding tension in one fell swoop. If map updates (which will still take place to inform flood risk management) indeed have no bearing on setting premiums, then there is no zone change that needs to be offset through grandfathering and a cross-subsidy.

Risk Rating 2.0 also promises another variety of 'equity' will be realised through this technical transformation. It will correct a flow that had the affluent paying too little, and ordinary people paying too much, by including the cost to rebuild in insurance rating. Under the current system, NFIP rates are set based on the dollar value of coverage and are not adjusted for the value of a home. What this means, for example, is that 'a $2 million dollar home may suffer $200,000 of loss more frequently than will a $200,000 home (for which it would be a total loss) but pay the same for coverage. Similarly, a flood that damages 10 percent of a building would cause more absolute damage for the higher value home, but again pricing is the same' (Kousky 2018, p. 25). In addition, the NFIP charges higher rates for coverage under a fixed dollar value, making insurance more expensive for lower valued homes that need less coverage.

In foregrounding the correction of this flow, FEMA is anticipating a common objection about the distributional effects of premium increases – and Risk Rating 2.0 is anticipated to increase premiums for most policyholders – namely, that policyholders' ability to absorb a price increase varies tremendously. For some, flood insurance increases mean their second vacation home on the beach becomes more expensive. For others, those increases put them in danger of not being able to maintain the mortgage on their only home. The objection that a shift to actuarial rates compounds existing socio-economic inequalities has led Congress to backtrack on earlier rounds of NFIP reform (Elliott 2017). Here again, Risk Rating 2.0 offers a technical way out of a persistent dilemma. If the algorithm can be tweaked, fine-tuned to take into consideration not only the risks people face, but also what they have of (financial) value, then the program can in a sense now 'see' these meaningful differences between policyholders and treat them accordingly.

In general, then, 'Risk Rating 2.0 is expected to lead to the reduction of cross-subsidies between NFIP policyholders, and the eventual elimination of premium subsidies and cross-subsidies once all properties are paying the full risk-based rate' (Horn 2021, p. 13). However, FEMA has indicated that it does not want to abandon at least one kind of cross-subsidy: that which funds a voluntary program called the Community Rating System (CRS). The CRS discounts insurance rates in communities that take steps to manage their flood risk; this can involve everything from simply advising people about flood hazards, to retrofitting flood-prone structures, to maintaining levees and dams. CRS discounts are offset by adjusting all premiums upward (Horn 2021). It is not clear how Risk Rating 2.0 will affect the CRS cross-subsidy, but FEMA's documentation continues to encourage participation in the program, to maintain a flow that rewards communities that go above and beyond the minimum standards.

What we have then, in this vision for the NFIP's future, is a flood insurance system that is capable of further individuating policyholders in multiple ways. This individualisation is normatively desirable for its ability to allocate to policyholders what they deserve – in terms of their assessed risk but also in terms of what they have to lose.

The United Kingdom: Flood Re and the political achievement of prudence

As in the NFIP, the existing arrangements for providing flood insurance in the United Kingdom involve an aspirational project of eliminating cross-subsidisation so that everyone is paying a risk-based rate. But whereas Risk Rating 2.0 promises to automate this change rather immediately as soon as a technical transformation to risk assessment takes place, in the United Kingdom, this change is projected to unfold more gradually and through the terms of a political agreement between the government and the private insurance industry, one which will eventually realise a 'free market' for flood insurance as everyone acclimatises to new kinds of responsibility.

Aspirations to gradually eliminate cross-subsidisation are expressed in the design of Flood Re, an officially temporary reinsurance arrangement launched in 2016, as a collaboration between the UK government and the insurance industry, that allows private insurers to continue to provide affordable flood insurance in high-risk areas. It works by allowing insurers to price policies at below-risk levels in such areas and cede the flood portion to Flood Re, which will reimburse any claims from a not-for-profit pool. The pool is funded by a levy on all insurers according to their market share, which they pass onto policyholders at an estimated £10.50 per policy (Surminski 2017; Penning-Rowsell 2015). The arrangement, 'scaffolded by subsidies,' is temporary insofar as it is regarded as 'a putative waystation to a free market' (Christophers 2019, pp. 5, 13). Indeed, as Surminski (2017) notes, a 'key justification for Flood Re's political approval despite its costliness was the argument that Flood Re does not create new forms of subsidisation, but merely formalises the already existing degree of cross-subsidisation' (p. 27). Flood Re is meant to operate for only 25 years. The 2018 update to the 'transition plan' indicates 'the subsidy required to deliver [affordable household insurance] cannot continue forever. The scheme was always intended to be time-limited and by 2039, Flood Re will have exited the market. When that happens, it is necessary that a market is in existence that is based both on risk-reflective pricing of household insurance and is affordable and available for households at risk of flooding' (Flood Re 2018, p. 8). Over time, the transition to risk-based pricing should take place commensurate with a kind of 'weaning' off subsidies (Christophers 2019, p. 17).

Why must this weaning take place? Why is it 'necessary' that this free market come into existence? Cross-subsidisation has always been a part of flood insurance in the United Kingdom, and it has not been terribly controversial. As Penning-Rowsell (2015) observes, 'the majority of the population is not discontent in cross-subsidising those unfortunate to live in areas liable to serious flooding' perhaps because they do not realise they are the source of the subsidy (p. 607). The choice of the word 'unfortunate' is also telling in this context. Exposure to flooding may be more commonly perceived as a function of bad luck, where residents are blameless victims who deserve compassion, not moral judgment or market discipline.

Cross-subsidies of the kind formalised under Flood Re are regarded as a flow in need of elimination not principally as a question of fairness, but rather as one of prudence. When low-risk policyholders offset the costs of high-risk policyholders, this dulls the purported incentive effects of insurance, where the price of premiums acts as a 'signal' of the underlying risk (Surminski & Eldridge 2017), as *The Economist* worried. This ostensibly perpetuates and even exacerbates the problem of high flood losses, in a vision where individuals are imagined to exert meaningful control over their exposure. With risk-reflective pricing, by contrast, policyholders can be made prudent – meaning 'rational, responsible, knowledgeable and calculative' (O'Malley 1996, p. 203; Christophers 2019). More prudent individuals, conscientious about their individual risk exposure, will undertake wiser decisions about where or how to live in relation to the water, proactively managing the present with an eye towards the future.

Again, such thinking is also at the core of the NFIP's commitment to actuarial rating (Collier 2014; Elliott 2021). But in the UK case, unlike in the United States, the aspiration to remove cross-subsidies is explicitly tied up in a project of minimising state activity in flood insurance. This project is one of pursuing what Christophers (2019) calls an 'allusive market': an asserted, but largely unspecified, free market for flood insurance that will be 'better' than Flood Re, the transitional solution in place today. In that better future, the government does not intervene directly to preserve affordability by setting rates (as in the United States) or by creating the conditions for cross-subsidisation to persist. The price of flood insurance is more 'appropriately' determined in a competitive market. The government *is*, however, responsible for creating the market conditions under which private insurers can continue operating profitably: namely, by protecting and even growing the insurable pool through '[f]uture government spending commitments on flood defences,' through 'improvements' to flooded properties, and to decreased building in floodplains (Flood Re 2018, p. 6). The elimination of the flow between policyholders is in some ways a by-product of making the free market possible. Even if policyholders themselves are not particularly bothered by cross-subsidising each other, for them to do so requires ongoing relations between the state and the market that are somehow inappropriate and that must be revised, so that the market can provide the solution, even where market solutions have thus far not been able to deliver on public aims of providing affordable and accessible coverage for all.

Whereas the existing flow between policyholders in the United States is rendered in itself unfair, in the United Kingdom the flow is problematic more for *what it seems to make possible*, in terms of how individuals, the government, and private insurers relate to flood risk and to each other. The *gradual* elimination of cross-subsidisation will give all stakeholders the runway they need to learn and develop the kinds of control and responsibilities necessary to achieve the more prudent, and market-led, management of flood risk.

Flows and overflows

In the United States and the United Kingdom, we find different strategies, a technical transformation and a political agreement, that both pursue future arrangements in which stigmatised flows between policyholders can be eliminated. In this future, where water does or is predicted to overflow riverbanks and beaches, to flow over the built environment, individual responsibilities stop at the property line. People mutualise the burden by participating in the risk pool of insurance, but their participation is tightly delimited by the portion of risk for which they can be made accountable. That this is so is a matter of equity and a matter of prudence.

That this will indeed *be* so is, however, by no means clear. The roll-out of Risk Rating 2.0 has already been slowed down and deferred, following concern from policyholders and their elected politicians about what will amount, for most people, to increases in their rates. FEMA promises the new rating system is coming, and it does not need Congressional authority to put it into place, because it is a technical adjustment rather than a statutory one. But the form Risk Rating 2.0 ultimately takes – how faithful it ultimately is to this initial vision (will it become 3.0, or 4.0, or…) – may well be circumscribed by claims from various stakeholders about what 'fairness' indeed means or requires when it comes to dealing with flood risk and loss. And in the United Kingdom, Flood Re is short on both details and on actual power to direct government agencies to deal with the underlying problem of flood risk. New home construction continues to take place on flood-prone land. What's more, 'The spatial shift in flood risk areas as a result of climate change is expected to disproportionately impact homes in deprived areas, notably in multicultural urban neighbourhoods and areas dominated by increasingly struggling homeowners' (Rözer & Surminski 2020) – in other words, risks are expected to increase more for people who are least able to afford a risk-based premium. Christophers (2019) thus reasonably predicts that, 'Precisely because a meaningful long-term strategy to address the underlying problems of flood risk management has not been developed now, those problems will continue to mount. In 2039, the necessary decisions will be tougher, and the necessary actions even more difficult' (p. 23). People will still face high flood risk and they will still be unable to afford a market price for flood insurance.

However, *even if* these aspirations are realised, even in partial form, the 'problem' of flow persists – it is just displaced. If risk-based premiums become unaffordable, people will not purchase flood insurance where they can avoid doing so (because mandatory purchase requirements don't exist or apply, or aren't widely enforced). Disaster aid, paid out of public coffers, still goes to help flooded areas rebuild. That flow comes from taxpayers, who send resources to recoup what are then even higher uninsured losses. Other flows, in the form of donations from charitable organisations and non-profits, also go to those in need. And recurrent issues of equity and prudent land use are anyway not vanquished by the elimination of cross-subsidies in flood insurance, through these or any other plans. Even if algorithms can discern the haves from the have-mores (the have-nots – the people who cannot acquire property – are entirely excluded), the pressures of risk-based pricing will unevenly affect policyholders, who have different resources available to take on mitigating action or to absorb any consequent hit to their property values. Even if a free market with unsubsidised and competitively derived insurance rates 'signals' risk, the pressures to build new housing and foster local economic growth, as well as the interventions of the extremely powerful real estate and finance industries in both countries, will throw up continued dilemmas related to where or whether to build, with more and more areas facing intensifying flood risks due in part to further climate change.

The elimination of the cross-subsidy flow could also lead to the creation of new flows. Where catastrophic losses bankrupt private insurers or lead them to 'defensively underwrite' and drop policies in the riskiest areas, we should expect some demand for new or expanded public backstops (indeed, this has taken place in the American West following the catastrophic wildfire seasons of the last few years). And where high risks and high premiums threaten property values, as well as community tax bases reliant on value-assessed property (typical of the United States in particular), then new forms of economic insecurity are created for individuals and communities, which may lead to demands for a tighter weaving of the social safety net or for flows of resources that make it possible to fortify the landscape with structural flood protection. The collective character of flood risk and its effects cannot be designed away in an algorithm or written out of the social contract between states and citizens.

The aspiration to eliminate cross-subsidies from flood insurance expresses a fantasy that we *can* be extricated from the encumbrances of each other and that, if we were, we would be empowered to manage on our own. Hanging onto the aspiration, relentlessly seeking it through new iterations of technical and political strategies, prevents us from considering that we might deal better with the coming floods by leaning into our interdependence, by expanding our forms of 'collective mutuality,' than we will by denying it or regarding it as only temporarily tolerable.

References

Anonymous 2014, 8 March, 'Waves of problems', *The Economist* 410, 8877, p. 76.

Baker, T 1996, 'On the genealogy of moral hazard', *Texas Law Review*, vol. 75, no. 2, pp. 237–292.

Baker, T 2002, 'Containing the promise of insurance: Adverse selection and risk classification', *Connecticut Insurance Law Journal*, vol. 9, no. 2, pp. 371–396.

Baker, T & Simon, J eds. 2002, *Embracing risk: The changing culture of insurance and responsibility*, Chicago: University of Chicago Press.

Christophers, B 2019, 'The allusive market: Insurance of flood risk in neoliberal Britain', *Economy and Society*, vol. 48, no. 1, pp. 1–29.

Collier, S.J 2014, 'Neoliberalism and natural disaster: Insurance as a political technology of catastrophe', *Journal of Cultural Economy*, vol. 7, no. 3, pp. 273–290.

Committee on the Affordability of National Flood Insurance Program Premiums 2015, *Affordability of National Flood Insurance Program Premiums – Report 1*. Washington, DC: National Academies Press.

Elliott, R 2017, 'Who pays for the next wave? The American welfare state and responsibility for flood risk', *Politics & Society*, vol. 45, no. 3, pp. 415–440.

Elliott, R 2021, *Underwater: Loss, flood insurance, and the moral economy of climate change in the United States*, New York: Columbia University Press.

FEMA 2021a, *Risk Rating 2.0: Equity in Action*, viewed 29 April 2021, <https://www.fema.gov/flood-insurance/work-with-nfip/risk-rating#:~:text=Risk%20Rating%202.0%20enables%20FEMA,and%20decreases%20are%20both%20equitable.&text=Because%20Risk%20Rating%202.0%20considers,a%20property's%20unique%20flood%20risk>.

FEMA 2021b, *Risk Rating 2.0 is Equity in Action: Fact Sheet (April 2021)*, Federal Emergency Management Agency, viewed 23 April 2021, <https://www.fema.gov/sites/default/files/documents/fema_rr-2.0-equity-action_0.pdf>.

Flood Re 2018, *Our Vision: Securing a Future of Affordable Flood Insurance*, <https://www.floodre.co.uk/wp-content/uploads/2018/07/Flood_Transition2018_AW.pdf>.

Horn, D.P 2021, 'National Flood Insurance Program: The Current Rating Structure and Risk Rating 2.0.' *Congressional Research Service*, R45999.

Kiviat, B 2019, 'The moral limits of predictive practices: The case of credit-based insurance scores', *American Sociological Review*, vol. 84, no. 6, pp. 1134–1158.

Koslov, Liz 2019, 'How maps make time: The temporal politics of life in the flood zone', *CITY*, vol. 23, no. 4–5, pp. 658–672.

Kousky, C 2018, 'Financing flood losses: A discussion of the national flood insurance program', *Risk Management and Insurance Review*, vol. 21, no. 1, pp. 11–32.

Kousky, C & Shabman, L 2014 (October), 'Pricing flood insurance: How and why the NFIP differs from a private insurance company', *Resources for the Future*, Discussion Paper 14–37, pp. 1–17.

Lamond, J & Penning-Rowsell, E 2014, 'The robustness of flood insurance regimes given changing risk resulting from climate change', *Climate Risk Management*, vol. 2, pp. 1–10.

Landes, X 2015, 'How fair is actuarial fairness?' *Journal of Business Ethics*, vol. 128, no. 3, pp. 519–533.

Lehtonen, T.K & Liukko, J 2015, 'Producing solidarity, inequality, and exclusion through insurance', *Res Publica*, vol. 21, pp. 1–15.

O'Malley, P 1996, 'Risk and responsibility', pp. 189–207 in *Foucault and political reason: Liberalism, neo-liberalism and rationalities of government*, Barry, A, Osborne, T & Rose, N (eds.), London: University College London Press.

O'Neill, J & Martin O'Neill 2012, *Social justice and the future of flood insurance*, York, UK: Joseph Rowntree Foundation.

Penning-Rowsell, E 2015, 'Flood insurance in the UK: A critical perspective.' *WIREs Water*, vol. 2, no. 6, pp. 601–608.

Rözer, V & Surminski, S 2020, 'New Build Homes, Flood Resilience and Environmental Justice – Current and Future Trends Under Climate Change across England and Wales', Centre for Climate Change Economics and Policy Working Paper No. 381.

Stone, D 2002, 'Beyond moral hazard: Insurance as moral opportunity', pp. 52–79 in *Embracing risk: The changing culture of insurance and responsibility*, Baker, T & Simon, J (eds.), Chicago: University of Chicago Press.

Surminski, S 2017, 'Fit for Purpose and Fit for the Future? An Evaluation of the UK's New Flood Reinsurance Pool', *Resources for the Future* Discussion Paper 17-04.

Surminski, S & Eldridge, J 2017, 'Flood insurance in England – An assessment of the current and newly proposed insurance scheme in the context of rising flood risk', *Journal of Flood Risk Management*, vol. 10, no. 4, pp. 415–435.

6 After the flood

Diverse discourses of resilience in the United States and Australia

Chloe Lucas and Travis Young

Introduction

Home insurance is seen as a central mechanism for disaster resilience by governments around the world (Booth & Tranter 2017; Lucas & Booth 2020). The 'empowerment' of individuals to manage and take responsibility for the risk of natural disasters impacting on their person and property is written into national and international strategies (e.g. United Nations Office for Disaster Risk Reduction 2009; Council of Australian Governments 2011). This reflects, on the one hand, a recognition of individuals' rights to participate in managing environmental risks that affect them; and on the other, a shift towards the responsibilisation of individuals, through which collective problems are reframed as the product of personal choices which can be best managed through market mechanisms (McLennan & Handmer 2012; Box et al. 2016).

Resilience (Holling 1973) is a contested, while ubiquitous term. Much of the literature on resilience has taken a systems approach that promotes a return to overall equilibrium following a shock, while allowing for internal adaptability (Walker et al. 2004; Folke et al. 2010). However, White and O'Hare (2014) argue that in policy and practice, resilience is most often used in a technocratic way – to measure the ability to bounce-back to an assumed non-threatening normality of business-as-usual. This is reflected in the definitions of resilience used by the insurance industry. For example, the Insurance Council of Australia (ICA) measures resilience as enabling the least disruption to 'business and individual capacity to operate normally' (ICA 2008, p. 3). In these insurantial rationalities, resilient individuals are seen to be self-reliant, self-disciplined, and privately insured. In this, the idea of resilience echoes a conservative ethos of non-interventionist, small government, and individual freedom to succeed or fail on one's own merits (Davoudi et al. 2012).

However, in a changed climate, normality is no longer non-threatening. Sea level rise and greater storm intensity means that homes are flooding with increasing frequency, and coastal property is at greater risk from storm surge (Thomas & Leichenko 2011; Christensen et al. 2007). Floods

DOI: 10.4324/9781003157571-8

are hard to mitigate at a property level, and mitigation efforts often involve major engineering, or retreat. For many people, after a flood there is no return to business-as-usual, just a growing realisation that their property will be subject to successive flood events, will lose its resale value, and that flood insurance may become unaffordable or unavailable. Insurance that does not enable transformative change to business-as-usual through adaptation to new normal climate realities can therefore be seen as impeding resilience (O'Hare et al. 2016; de Vet et al. 2019). Community understandings and lived experiences of resilience may also be different to – and potentially at odds with – insurantial rationalities (Wamsler & Lawson 2011; Adger et al. 2013)

In this chapter, we describe cases of flood and storm damage that occurred in Hobart, Australia (2018), and Houston, United States (2017). Using these case studies, we examine the ways in which flood resilience is discursively and structurally constructed by governments, insurers, and the public. While these floods took place in very different parts of the world, with different landscapes, insurance systems and social contexts, we found some overlap between Hobart and Houston residents' experiences of flood insurance, and hence their respective discourses. We investigate how the experience of insurance (or lack of insurance) affected the resilience discourse of householders whose homes were flooded. We identify a technocratic discourse of resilience perpetuated by insurers and governments, which focusses on financial recovery and bounce-back to equilibrium. Insured householders, however, described their resilience as affected by the uncertainty created through the insurance experience, and dependent on their capacity to advocate for their rights in a challenging insurance bureaucracy. The insurance-driven recovery process exposed an uneven landscape of vulnerability, access to resources, and political power – illustrating insurance's duality in expediting recovery and amplifying existing disparities. We find that contrary to the insurance discourses of government and industry, community resilience is more than financial. We suggest that both insurers and governments should be mindful of the need for equity and transparency in order to maintain the social legitimacy of flood insurance.

Weather extremes and insurance in Houston and Hobart

While both are located in western liberal countries, our case studies are in cities with very different geography, history, development, and riskscapes. These cities are different in their population and built environment. Hobart has a largely middle-class, white population of less than 250,000, living in low-density housing, often close to natural features including forest, creeks, and coastline. Houston is a massive, sprawling city developed in a low-lying coastal area. It is one of the most diverse cities in the United States, and its 2.3 million people live in a mix of low- and high-income neighbourhoods, often segregated along lines of race and ethnicity.

Greater Hobart is the capital of the Australian island state of Tasmania. Although it has a temperate climate, Tasmania is located in the 'Roaring forties' – the strong prevailing winds of the southern hemisphere – and its weather is famously changeable. Hobart is at high risk of bushfire, and its position at the foot of kunanyi/Mount Wellington makes it prone to flash flooding.

Greater Houston is at the forefront of urbanisation and climate change. The region has been devastated by storm and flood events for over a century, including the deadliest hurricane in US history – the Galveston Storm of 1900. More recently, the metropolitan region has been impacted by Tropical Storm Allison (2001), Hurricane Ike (2008), and major floods in 2015 and 2016. In August 2017, Hurricane Harvey soaked the region, bringing record rainfall and property damage.

The United States and Australia also have different insurance contexts. Flooding is the costliest disaster in the United States, with damages increasing significantly in the last few decades (Thomas & Leichenko 2011). In the first half of the 20th century, private insurers left the market as flood losses mounted across the United States (Knowles & Kunreuther 2014). In 1968, the United States government created the National Flood Insurance Program (NFIP) to subsidise premiums for homeowners and incentivise community-level investment in flood management – helping communities and households to bounce-back faster after a flood (Thomas & Leichenko 2011).

Australia is subject to periodic flooding both from storm-related events and cyclones, and from catchment-wide rains across extensive floodplains. As a continent it has the greatest annual variability in rainfall and runoff, and often experiences long periods of drought, so floods can come as a surprise (Wenger et al. 2013). Australia has long taken a market-led approach to disaster insurance, and flood cover has historically been included in a minority of household insurance policies. In some parts of the country, including large parts of northern Australia, flood insurance is either unavailable or unaffordable. This reflects an industry standard to price insurance according to localised risk data, as well as higher numbers of extreme events and high costs for operating in remote environments (ICA 2018a). Non-insurance and underinsurance are widespread in Australia, in part because of high levels of public mistrust of insurers (Tranter & Booth 2019) and because of the trade-offs undertaken by households in their insurance decision-making (Booth & Harwood 2016).

These differences between Hobart and Houston's landscape, community and insurance contexts generate distinct lived experiences of flood insurance, and thus case studies that complement one another, as well overlapping in some findings.

Hobart: Managing insurantial uncertainty

An extreme weather event on 11 May 2018 brought record-breaking rainfall of 129 mm in 24 hours, high winds, and flooding to Hobart. Almost

5000 house and 3000 contents insurance claims were made, totalling almost AU $100 million (ICA 2018b). We interviewed 15 residents of Greater Hobart whose homes had been damaged in the event. Interviewees were six men and nine women, across a wide age range from their early 20s to mid-80s. Interviews occurred between one month and four months after the flood event. Three of the participants were interviewed multiple times, at approximately monthly interviews from one month to four months after the event, to document their ongoing experience after the flood.

Thirteen participants were homeowners, two were renters. Most of their homes had been damaged by pluvial floodwater overwhelming the storm-water system, but one was also damaged by a fluvial flood, and one had its roof blown off in the storm. Twelve of the thirteen homeowners put in insurance claims, while neither of the renters were covered by insurance, and one homeowner living in the floodplain decided not to claim for damage to their property.

Among the insured participants, the experience of negotiating insurance claims was broadly similar. This experience was notable for the way in which insurance rationalities acted to contradict the knowledge and testimony of the insured, and to destabilise participants' sense of control and agency in the aftermath of the disaster. Participants described their attempts to navigate insurance jargon, worried that using the wrong terms 'might give [the insurer] little loopholes,' and reading the policy document 'with a fine-toothed comb' to be prepared with the right (insurance) language to negotiate (Yvette, homeowner, 40s). This reflects participants' partial engagement with insurance discourses around flood, which differentiate between stormwater inundation and water escaped from watercourses such as rivers. Participants were concerned that some forms of flood might be excluded, but, as Herman (homeowner, 50s) described, in practice it made little difference to the experience: 'The stormwater washed things into the floodwater... Well, it was all bloody rain.'

They described the arrival of multiple assessors with diverse claims of expertise, from hydrologists and hygienists to plumbers and loss adjustors, come to document the type of damage, and how it might be covered by the insurance. Insured participants all described this as a disorienting process because of the number and variety of experts sent by the insurers to assess cases – often more than ten for one property. In most cases, these actors produced reports that either contradicted householders' own testimony, or contradicted one another.

> One group would come and say, 'yes, your curtains need replacing because of the mould and the smell'. One would say 'the whole bed base and the mattress all gets replaced'. The next ones would say, 'no, just the mattress. We don't replace the base'. So we didn't really know where we were.
>
> (Rowan, homeowner, 80s)

Five of the twelve participants who put in claims described calling their insurers to discuss these assessments and being told there were no records of them. While this could be put down to insurers' systems being overwhelmed by an extraordinary event, it should be noted that these kinds of events are exactly what insurers claim to be prepared for. The Insurance Council of Australia's 'Understand insurance' website, for example, reassures customers that:

> Individual insurance companies are accustomed to meeting the claims demand placed upon them both during normal business and times of crisis.
>
> (ICA, undated)

In practice, this was not the case in Hobart, and other examples suggest that in times of disaster, failures in the insurance claim process may be the norm (e.g. ACCC 2019). This does insurers themselves little harm – faced with confusion and uncertainty, customers may be willing to settle for smaller claims, or take a cash payout rather than a managed program of works.

> They lost our receipts that we'd sent in three times saying, 'We're carrying this $2600 debt and we can't afford it.' When I finally contacted them they said they had no record of any of those communications, no record that there's any internal damage, this hole was in the ceiling, all the doorframes were swollen, stains on the wall.
>
> (Beryl, homeowner, 40s)

The uncertainty of outcome and lack of transparency over how assessments were made, as well as multiple changing insurance experts who could not be pinned down, and the intractable process of getting repairs or cash settlements out of insurers had the effect of limiting participants' resilience. This was particularly hard for participants who had limited capacity because of age or disability. Beryl, who lives with brain injury, described how, four months after the event, it was still taking a toll on her everyday life:

> It's a shambles. And because there's stains everywhere and people are still coming to do the assessment I've given up on cleaning... because you feel like you're living in a dive... And it does take energy to go 'Oh, that's not resolved yet. I have to call them again.' You know I spent two hours the other day just explaining my story again and it takes energy. So, no, we're not recovered.
>
> (Beryl)

Participants described their resilience as reliant on their ability to manage insurance rationalities and practices. This meant being constantly available, flexible, and patient; but also assertive and knowledgeable, proactive and energetic, and able to push-back when offered a lesser outcome than they felt entitled to. Keeping their own records was thought to be important – one interviewee described having compiled a database of more than 150 inter-actions with insurers or their representatives over the two months since the event. In a different take on the normative discourse of being good *because* you are insured, participants felt their personal capacity and qualities ena-bled them to be good *at being* insured:

> I feel like it takes a lot of assertiveness and persistence to make sure I got what we deserved and needed.
>
> (Yvette)

In this way, insurance contributed to exacerbating differences in resilience between people who had the time, money, and capacity to manage the insur-ance claims process, and people living with existing disadvantage.

Houston: Insurance entwined with inequality

After making landfall on 25 August 2017, Hurricane Harvey made its way north to Houston where it sat over the city for days. Harvey produced over 1,270 mm of rain, damaged over 200,000 homes, and caused over US $125 billion in total damages (Blake & Zelinsky 2018). Over US $9 billion was paid out through the NFIP in the Houston area, spread between 92,000 insurance claims (US FEMA 2019).

During the nine months following Harvey, we interviewed 38 renters and homeowners in two Houston neighbourhoods[1]. While physically close, the neighbourhoods were distinct in their demographic and socioeconomic make-up: (1) historically Black, lower-income, with equal renter and home-ownership rates; (2) majority white, upper-income, and higher renter rates. While both communities were impacted by Hurricane Harvey, the former received significantly more damage. Of the interviewees, half the homeown-ers had flood insurance and nearly all had experience with flood insurance in the last 15 years. Only one renter had flood insurance, but over half had some form of contents insurance. The interviewees' experiences of flooding and insurance were markedly different depending on socioeconomic status and knowledge of the insurance system.

While flood insurance has been promoted as a key mechanism in sup-porting resilience, disparities in insurance and recovery experiences reflect embedded economic and racial inequities in Houston and other US metros (Collins et al. 2019; Paganini 2019; Grove et al. 2020). Low-income families struggle to find affordable flood insurance, so when a flood damages their home, they must rely on limited savings, charity, and/or government dis-aster relief. However, these cannot be counted on to arrive at all, let alone provide a sufficient amount of support to recover.

Renters and homeowners have different experiences with insurance during disaster recovery that leave renters disproportionately disadvantaged (Emrich et al. 2020). In the United States, standard contents insurance does not cover flood damage, nor does the property owner's flood insurance cover a renter's belongings. The NFIP provides subsidised rental coverage, but renters are rarely aware of this; nor do landlords or leasing agents readily disclose flood history or status. While Karen (renter, 40s) escaped Harvey with minimal property damage, her past experiences are common of many Houston renters:

[KAREN] You know that flood insurance doesn't cover renters.

[INTERVIEWER] What do you mean?

[KAREN] I mean…we got flooded in 2015. The property owner had flood insurance and got money for repairs, but I didn't get any help…I thought I would be covered under his policy.

[INTERVIEWER] What did you do?

[KAREN] I was lucky. Our lease was ending, so we weren't on the hook for any rent….But our neighbours ended up paying [or being charged] for almost six months [without being able to inhabit].

As Karen described, a lack of knowledge about flood exposure and insurance can have repercussions beyond damage to personal property. In this sense, a vulnerable situation is compounded when socioeconomic and bureaucratic mechanisms further exacerbate disparities between renters and property owners. While renters are scrambling for temporary accommodation, insured property owners can use their insurance to rebuild and upgrade properties, pricing out previous renters. A flood event that damages the resilience of some renters can simultaneously increase the resilience of some property owners. The benefits realised by landlords are passed on to local governments in the form of tax revenue, and supported under the banner of a more resilient community.

Resilience discourses at the community level often include broad messages of building-back-better to maximise economic benefits. However, in practice, these benefits are realised by the few at the expense of the many. One Houston community advocate described resilience messaging and its associated development after a disaster:

It's gentrification… instead of fancy stores and big houses pushing people out it's a hurricane or a storm. After that, the city and the developers see opportunity.

(Raymond)

These sweeping changes are particularly acute in lower-income areas with older, long-term homeowners.

NFIP payouts amount to less and are fewer in number in lower-income areas and in communities with a higher percentage of racial and ethnic minorities (Emrich et al. 2020). These deficits increase as social vulnerabilities accumulate. This means that lower-income renters and homeowners, particularly people of colour, are more likely to be permanently displaced; and if they do return, they are met with a disjointed sociocultural landscape.

While previous experience with insurance aids some, in many cases, that previous experience reinforces disparities in access to more resilient recovery. Regina (homeowner, 50s) had an unfavourable experience with flood insurance in the past. That experience, along with increasing premiums, led her to decide not to purchase insurance despite recurring flood losses:

[REGINA] We had insurance. We had it during [Hurricane] Ike and it wasn't worth it… The back and forth [with the insurance company] was one big headache… We ended up only getting a few thousand dollars because they said most of the damage was caused by wind.
[INTERVIEWER] Have you been flooded since then?
[REGINA] Yes, a couple of times…We're in the 100 [100-year floodplain].
[INTERVIEWER] Is flood insurance required?
[REGINA] Not for us, but yes… this was my parent's house. When there was a mortgage, we had it. But once it was paid off, we didn't have to have it…and it's a lot more expensive now!… so now we just rehab a little bit at a time. We might get help from the church or my uncle.

Regina's experience contrasts with that of Ken (homeowner, 40s), who lived 5 km to the south:

[KEN] I have flood insurance, but I'm not in the floodplain. It's really affordable, so I thought 'why not.'
[INTERVIEWER] Were you flooded during Harvey?
[KEN] No. Water came up in the yard, but it never reached the house… I moved up here [Houston] from Galveston a couple of years ago… my house was flooded during Ike. I made off a lot better than a lot of my neighbours… both my parents worked in insurance, so I knew what to do.

While government and industry discourses reason that insurance coverage equals more resilience, that messaging is not evenly experienced. Ken was able to use his social capital and networks to navigate the system and increase his resilience – making the system 'work for him.' However, for Regina, resilience was found in eschewing flood insurance and relying on self-repairs, networking, and informal recovery means. Ken's success was reinforced by direct insurance benefits and a 'more resilient' housing market spurred on by recovery investment – allowing him to profit when selling his previous residence. Regina's recovery was counter to broader resilience and redevelopment discourses, meaning her recovery (and resilience) centred on relationships and sense of community.

How is flood resilience discursively constructed by insurers and governments in the United States and Australia?

Resilience discourse has not only dominated disaster research in recent decades, but has driven recovery policies with complicated results (Tierney 2015). Evolving interpretations of resilience have become integral in the ways governments, the insurance industry, and individuals navigate the hazard landscape. Driven by an array of actors, resilience discourses are (re)framed and (re)produced across scales – driving major flood infrastructure projects, incentivising development, and individualising risk.

Our case studies illustrate the various roles that government and private interests play in constructing resilience, assigning responsibility, and shaping socioeconomic futures. Resilience discourses have become 'usefully ambiguous,' as their fluidity and malleability allow different actors to use them in different ways at different times (Tierney 2015). Often, this ambiguity allows for symbiotic partnerships that reduce insurance exposure and allow for development in vulnerable and previously undevelopable areas. Through lobbying efforts (including withdrawal of insurance), the insurance industry convinces the government that the construction of reservoir and levee projects increases the resilience of surrounding communities to impacts from flooding. This allows for increased development (broader tax base) while protecting private interests (flood insurance liability). This pattern of events has occurred multiple times in the United States (Kunreuther 2006) and Australia (McAneney et al. 2016).

Relationships between governments and the insurance industry are instrumental in the way resilience is constructed. In the United States, the dominant discourse of the NFIP is that it makes communities more resilient to the impacts of flooding by easing the financial burden on homeowners and creating a baseline for flood management. In this sense, resilience is constructed as a shared goal, one in which private insurers, governments, and homeowners work together to ensure that economic baselines – such as tax revenue and development incentives – are maintained. This shared goal makes sure the 'business' in business-as-usual is front and centre.

However, the lived experience of households impacted by flooding can be complicated by the NFIP. In particular, the program can incentivise development in high-risk areas, with long-term negative consequences for homeowners (Kunreuther 2006). Outdated flood management (mapping, building codes, and infrastructure) provides a false sense of security to residents in exposed areas, and the subsidisation of insurance masks the premiums needed to make and keep the NFIP solvent (Kunreuther 2006).

In Australia, despite private insurance being central to government discourses of climate resilience (Lucas & Booth 2020) and no public flood/disaster insurance schemes, inadequate insurance is still prominent during flood disasters. Lack of flood insurance availability, affordability, and transparency has been the subject of multiple independent reviews (Mason 2011).

A strong and effective insurance lobby has been successful in opposing calls for government intervention and subsidisation (Dolk & Penning-Rowsell 2020). The Australian non-life insurance industry is the most profitable in the world (Dolk & Penning-Rowsell 2020), and is highly motivated to minimise government's role with regard to insurance (ICA 2011; Konrad & Thum 2012). Governments have been strongly influenced by the technical and actuarial discourse of resilience, and national and state disaster preparedness materials increasingly cite insurance knowledge, and use of insurance calculators as necessary for resilience, implying moral norms of 'good' levels of insurance (Booth & Harwood 2016).

More than financial resilience

Despite their very different insurance contexts, in both Australia and the United States, financial resilience via insurance is represented in government and insurance discourse as community resilience. As Weinkle (2019, p. 3) notes:

> Rhetoric surrounding the insurance industry... shifts attention away from the political activities embedded in technical practices, assuming these practices as separate from the social realm.

In many ways, Hobart and Houston are on opposite ends of the urban spectrum. However, their cases present complementary evidence of how different household insurance regimes reproduce community-level financial and technocratic resilience discourses while cloaking insurance decision-making and further entrenching disadvantage.

In both Hobart and Houston, post-flood resilience was experienced differently by those with differing levels of (dis)advantage. While being insured offered some financial resilience, insured participants' perception of their resilience was dependent on their capacity to manage uncertain and technocratic insurance discourses. Far from the advertised expectation that being insured allows one to hand over the problem (and the worry) to the experts, flood victims found that in practice, resilience involved keeping records of interactions with insurers, doing interim repairs themselves, and necessitated arguing for their entitlements. While participants with time, education, and negotiating experience were largely successful, this process further entrenched inequality as disadvantaged flood victims were persuaded to settle for less, or decided in the end to drop out of the insurance process.

Both cases provide examples of how flood insurance can reinforce resilience for some and undermine resilience for others. Renters found themselves outside of the flood insurance system – uninsured, underinsured, and/or unaware of their flood risk. Insurance-driven recovery processes left renters doubly disadvantaged, often paying for rent on an uninhabitable home while searching for new or temporary residence. In some cases, property owners utilised insurance coverage to update properties and increase rental income, turning temporary relocation to permanent displacement for some renters. This produced more financially resilient landlords and neighbourhoods, while increasingly vulnerable tenants carried a heavier financial burden without key social assets. In Houston, this meant that lower-income households were more easily displaced and replaced with people who better fit the new resilience aesthetic. Hobart's already middle-class neighbourhoods affected by the floods were at less risk of gentrification, but the Hobart case shows how other forms of disadvantage are exacerbated by insurance experiences.

For governments managing residential landscapes of high climatic risk, whether they be floodplains, areas prone to wildfire, erosion, heatwaves, or storms, it is increasingly important to develop adaptive and equitable forms of disaster resilience (Grove et al. 2020). It should not be assumed that insurance alone is sufficient for this complex and dynamic social, psychological, and material process. While supporting financial resilience is vital in mitigation and recovery settings, recognising the contextual elements of resilience, including householders' diverse capacities to manage the insurance process, is paramount in making insurantial experience and the broader recovery process more accessible and equitable. Insurers themselves should also consider the importance of transparency and fairness in ensuring their public legitimacy and social license into the future.

Note

1 Many of the interviews in the Houston case occurred during community-rebuilding events. Quotes are taken from field notes and are based on multiple conversations with affected homeowners and renters.

References

Adger, W.N, Quinn, T, Lorenzoni, I et al. 2013, 'Changing social contracts in climate-change adaptation', *Nature Climate Change*, vol. 3, pp. 330–333.

Australian Competition and Consumer Commission (ACCC) 2019, *Northern Australia Insurance Inquiry Second Update Report*, <https://www.accc.gov.au/focus-areas/inquiries-ongoing/northern-australia-insurance-inquiry/second-interim-report>.

Blake, E.S, & Zelinsky, D.A 2018, *National Hurricane Center Tropical Cyclone Report: Hurricane Harvey*, National Oceanic and Atmospheric Administration, <https://www.nhc.noaa.gov/data/tcr/AL092017_Harvey.pdf>.

Booth, K & Harwood, A 2016 'Insurance as catastrophe: A geography of house and contents insurance in bushfire-prone places', *Geoforum*, vol. 69, pp. 44–52.

Booth, K & Tranter, B 2017, 'When disaster strikes: Under-insurance in Australian households', *Urban Studies*, vol. 55, no. 14, pp. 3135–3150.

Box P, Bird D, Haynes K, King D 2016, 'Shared responsibility and social vulnerability in the 2011 Brisbane flood', *Natural Hazards*, vol. 81, pp. 1549–1568.

Collins, T.W, Grineski, S.E, Chakraborty, J & Flores, A.B 2019, 'Environmental injustice and Hurricane Harvey: A household-level study of socially disparate flood exposures in Greater Houston, Texas, USA', *Environmental Research*, vol. 179, 108772.

Council of Australian Governments 2011, *National Strategy for Disaster Resilience*, <https://knowledge.aidr.org.au/resources/national-strategy-for-disaster-resilience/>.

Christensen, J, Hewitson, B & Busuioc, A et al. 2007, *Regional climate projections, climate change 2007: The physical science basis*, Contribution of Working group I to the Fourth Assessment Report of the Intergovernmental Panel on Climate Change, University Press, Cambridge.

Davoudi, S, Shaw K & Haider, L et al. 2012, 'Resilience: A bridging concept or a dead end?', *Planning Theory and Practice*, vol. 13, pp. 299–333.

de Vet, E, Eriksen, C, Booth, K & French, S 2019, 'An unmitigated disaster: Shifting from response and recovery to mitigation for an insurable future', *International Journal of Disaster Risk Science*, vol. 10, pp. 179–192.

Dolk, M & Penning-Rowsell, E.C 2020, 'Advocacy coalitions and flood insurance: Power and policies in the Australian Natural Disaster Insurance Review', *Environment and Planning C: Politics and Space*, vol. 39, no. 6, pp. 1172–1191.

Emrich, C, Tate, E, Larson, S.E & Zhou, Y 2020, 'Measuring social equity in flood recovery funding', *Environmental Hazards*, vol. 19, no. 3, pp. 228–250.

Folke, C, Carpenter, S & Walker, B et al. 2010, 'Resilience thinking: Integrating resilience, adaptability and transformability', *Ecology and Society*, vol. 15, no. 4, p. 20.

Grove, K, Barnett, A & Cox, S 2020, 'Designing justice? Race and the limits of recognition in greater Miami resilience planning', *Geoforum*, vol. 117, pp. 134–143.

Holling, C.S 1973, 'Resilience and stability of ecological systems', *Annual Review of Ecology, Evolution, and Systematics*, vol. 4, pp. 1–23.

Insurance Council of Australia (ICA) (undated) *Understanding Insurance*, <https://understandinsurance.com.au/recovering-from-a-disaster>.

Insurance Council of Australia (ICA) 2008, *2008 Federal Pre Budget Submission Improving Community Resilience to Extreme Weather Events*, <http://www.insurancecouncil.com.au/assets/files/community%20resilience%20policy%20150408.pdf>.

Insurance Council of Australia (ICA) 2011, *Response to 2011 Natural Disaster Insurance Review*, <https://treasury.gov.au/sites/default/files/2019-03/c2011-ndir-ip-Insurance_Council_of_Australia.pdf>.

Insurance Council of Australia (ICA) 2018a, *Response to ACCC Issues paper – Northern Australia Insurance Inquiry*, <https://www.accc.gov.au/system/files/Insurance Council of Australia %28ICA%29.pdf>.

Insurance Council of Australia (ICA) 2018b, *More than half of Hobart flood and storm claims closed as insurance bill nears $100m*, <https://www.insurancecouncil.com.au/assets/media_release/2018/200818%20More%20than%20half%20of%20Hobart%20flood%20and%20storm%20claims%20closed.pdf>.

Konrad, K.A & Thum, M 2012, *The role of economic policy in climate change adaptation*, EIB Working Papers, European Investment Bank, 2012 (02).

Knowles, S.G & Kunreuther, H 2014, 'Troubled waters: The national flood insurance program in historical perspective', *Journal of Policy History*, vol. 26, no. 3, pp. 327–353.

Kunreuther, H 2006, 'Disaster mitigation and insurance: Learning from Katrina', *The Annals of the American Academy of Political and Social Science*, vol. 604, no. 1, pp. 208–227.

Lucas, C.H & Booth, K 2020, 'Privatizing climate adaptation: How insurance weakens solidaristic and collective disaster recovery', *WIRES Climate Change*, vol. 11, no. 6, pp. 1757–7780.

Mason, A 2011, *The History of Flood Insurance in Australia*, Paper commissioned for the National Disaster Insurance Review, <https://treasury.gov.au/review/natural-disaster-insurance-review/commissioned-papers>.

McAneney, J, McAneney, D & Musulin, R et al. 2016, 'Government-sponsored natural disaster insurance pools: A view from down-under', *International Journal of Disaster Risk Reduction*, vol. 15, pp. 1–9.

McLennan, B.J & Handmer, J 2012, 'Reframing responsibility-sharing for bushfire risk management in Australia after Black Saturday', *Environmental Hazards*, vol. 11, pp. 1–15.

O'Hare, P, White, I & Connelly, A 2016, 'Insurance as maladaptation: Resilience and the 'business as usual' paradox', *Environment and Planning C: Government and Policy*, vol. 34, no. 6, pp. 1175–1193.

Paganini, Z 2019, 'Underwater: Resilience, racialised housing, and the national flood insurance program in Canarsie, Brooklyn', *Geoforum*, vol. 104, pp. 25–35.

Thomas, A & Leichenko, R 2011, 'Adaptation through insurance: Lessons from the NFIP', *International Journal of Climate Change Strategies and Management*, vol. 3, no. 3, pp. 250–263.

Tierney, K 2015, 'Resilience and the neoliberal project: Discourses, critiques, practices—and Katrina', *American Behavioral Scientist*, vol. 59, no. 10, pp. 1327–1342.

Tranter, B & Booth, K 2019, 'Geographies of trust: Socio-spatial variegations of trust in insurance', *Geoforum*, vol. 107, pp. 199–206.

United Nations Office for Disaster Risk Reduction 2009, Sendai Framework for Disaster Risk Reduction 2015-2030, <https://www.undrr.org/publication/sendai-framework-disaster-risk-reduction-2015-2030>.

US Federal Emergency Management Agency (FEMA) 2019, *Two years after Harvey, flood insurance is still a smart investment*, NR-120, < https://www.fema.gov/press-release/20210318/two-years-after-harvey-flood-insurance-still-smart-investment >.

Walker, B, Holling, C, Carpenter, S & Kinzig, A 2004, 'Resilience, adaptability and transformability in social-ecological systems', *Ecology and Society*, vol. 9, p. 5.

Wamsler, C & Lawson, N 2011, 'The role of formal and informal insurance mechanisms for reducing urban disaster risk: A South-North comparison', *Housing Studies*, vol. 26, pp. 197–223.

Weinkle, J 2019, 'Experts, regulatory capture, and the 'governor's dilemma': The politics of hurricane risk science and insurance', *Regulation and Governance*, vol. 14, no. 4, pp. 637–652.

Wenger, C, Hussey, K & Pittock, J 2013, *Living with floods: Key lessons from Australia and abroad*, National Climate Change Adaptation and Research Facility, Gold Coast.

White, I & O'Hare, P 2014, 'From rhetoric to reality: Which resilience, why resilience, and whose resilience in spatial planning?', *Environment and Planning C: Government and Policy*, vol. 32, pp. 934–950.

7 Flood insurance

A governance mechanism for supporting equitable risk reduction and adaptation?

Mark Kammerbauer and Christine Wamsler

Introduction

In Germany, increasing numbers and severity of flood events result in significant economic losses. Flood insurance is considered an important governance mechanism to reduce related long-term impacts, involving private market, governmental and civil society actors (Kammerbauer 2019). However, whilst flood impacts are increasing and flood insurance is available to all homeowners, coverage is low. Why do such disparities persist and how can they be addressed to reduce risks and to assist particularly vulnerable populations? How effective is the current governance system for managing and adapting to hazard and disaster impacts? How does it relate to efforts to adapt to climate change? And what role can insurance play to improve the current governance mechanisms for supporting equitable risk reduction and adaptation?

We explore these questions based on a case study of the German city of Deggendorf, which was strongly affected by the 2013 Danube flood. The case shows that being insured against floods is no guarantee for citizens to be able to 'build back better.' At the same time, it discloses different equity issues related to flood insurance. In this book chapter, we explain this situation and the associated contradictions and complexities of the recovery phase after the Danube flood. After introducing our case study context and methods, we describe the flood governance system and policies in Germany and how they interrelate with urban planning, climate change policies, and flood insurance. The following results section shows how the Danube flood played out in our case study: who was impacted and why, who had insurance coverage and who not, and who could therefore benefit (or was excluded) from financial governmental assistance to cover rebuilding costs. The results indicate discrepancies between the concept of civil protection in Germany and the EU-wide initiative for flood management with its emphasis on individual responsibility. In this context, insurance plays a counterproductive role as it has not led to the reduction of vulnerabilities of at-risk communities. We conclude with recommendations on how to improve the role of flood insurance for sustainable and equitable flood risk governance.

DOI: 10.4324/9781003157571-9

In this chapter, vulnerability is understood as 'the characteristics and circumstances of a community, system or asset that make it susceptible to the damaging effects of a hazard' (UNDRR). Disaster studies oriented on human ecology show that the outcomes of disasters are rooted in the everyday lives of those impacted by them (Susman et al. 1983). The confluence of historic root causes, dynamic factors, and contextual uncertainty can, in this context, be understood as the process that 'drives' vulnerability. One principal cause for vulnerability is the lack of access to resources (e.g. disaster insurance) required to cope with a disaster. This is influenced by variables including social status, occupation, ethnicity, gender, disabilities, health, age, and migration background (Wisner et al. 2004, p. 11). Viewed from this perspective, vulnerability can be brought into relation with community-wide social and political structures, while indicating to what extent households can cope with processes of change. Social networks are also crucial in this context as they can reduce vulnerability by providing related support (Chambers 1989; Bohle 2001, pp. 3–5). Vulnerability can therefore be understood as a dynamic process, which is embedded within the respective social, political, and economic context. Its root causes impair access to resources required for adequate means of risk reduction and adaptation to rapid onset or long-term change in the context of environmental risk.

Case study context and methods

During May and June 2013, strong and persistent precipitation of up to 300 per cent of the monthly average led to multiple floods in communities in northern Bavaria, eastern Germany as well as Czechia. As a result, a disaster was officially declared across eight German federal states and 80,000 residents receive an evacuation notice on 10 June 2013. The regional authority task force in Deggendorf coordinated the local response and 6,000 residents were notified to evacuate. Along the Danube and the Isar rivers, floodwaters rose up to 8 meters above normal levels. Two dikes breached along the Danube and the Isar, flooding the adjacent landscape. Floodwaters submerged the Fischerdorf and Natternberg areas of Deggendorf, reaching 2 meters in height.

While first responders tried to repair the breached dike, heating oil tanks and supply lines in Natternberg and Fischerdorf were ruptured by the flood. Oil merged with standing floodwaters for up to 18 days, penetrating deep into walls, contaminating buildings. In the Deggendorf region, nearly 1,000 houses were impacted, among them 600 in Fischerdorf and 90 in Natternberg (Kammerbauer & Kaltenbach 2014; Landratsamt Deggendorf 2017b).

The combination of floodwaters and heating oil led to an exceptional compound disaster that affected the population unevenly throughout the flooded area. In cooperation with the Rural Development Agency, publicly appointed expert surveyors were tasked with providing consultation to impacted homeowners. Due to the severity and depth of oil contamination of walls, floor slabs, and other building components, more homes than initially expected were uninhabitable and needed to be demolished.

The majority of homeowners were not insured against floods and became eligible to apply for governmental funding for rebuilding homes. For the administrative region of Deggendorf, 118 million euros was provided by late May 2017 and distributed to 893 applicants. Among these, 227 applied for demolition and 666 applied for rehabilitation (Deutscher Bundestag 2013; Landratsamt Deggendorf 2017a). At the same time, the German federal and state governments agreed to establish a program for immediate assistance to private households in order to provide funding for 80 per cent of damages to the homes of uninsured owners. Donations managed by a charitable board were supposed to cover the remaining costs. Applications were handled by the district administration and the city building department. The duration of the program was initially limited to three years, but the application process was lengthy and the submission deadline extended until late 2018. Needy homeowners were eligible to receive up to 100 per cent of costs. For applications to be successful, uninsured homeowners needed to demonstrate that the new structure was equivalent to the previous building (Kammerbauer & Wamsler 2018).

A case study design with a mixed methods approach (Gerring 2007, p. 90; Yin 2009, p. 18) was employed to examine this case in detail. Our data originate from site visits and follow-up enquiries that took place from September 2013 until the end of 2018. Methods included, first, a questionnaire survey featuring quantitative and qualitative questions, which was distributed among 53 households (of a statistical population of 374 households); second, a quantitative spatial analysis of official address books issued by the municipality and the flooded parts of the city; third, qualitative interviews with residents and experts; fourth, a qualitative content analysis and triangulation of data in order to test and validate results; and fifth, a discussion of preliminary results with the interviewed experts for feedback and verification.

The case study focuses on the households and buildings in the flood zone. The selection of the survey participants was purposeful (with assistance of a local contact person and by asking residents to participate) and employed the 'snowball method' (Schnell et al. 2011, p. 294). 374 residential addresses were identified as statistical population. The questionnaire survey (distributed in-person in May 2014 and collected on the same day and within the following week) covered 53 addresses, corresponding to the number of participating households, and involved 130 household members.

In the spatial analysis, the official address books of the municipality of Deggendorf served to substantiate the survey data by comparing addresses before (2012) and after (2014) the flood. The analysed differences help to identify changes in relation to the built environment and its residents. The address books list all extant addresses in combination with names of residents. 399 addresses were located in the flood zone. 374 addresses indicated the presence of at least one residential unit. 50

addresses featured three or more family names within a multi-family residential structure. 20 per cent of addresses listed in the 2012/2013 address book were no longer listed in the 2014/15 edition. The related buildings were either demolished or vacant. In addition, another 20 per cent showed strong fluctuation among residents.

The interview participants were also purposefully selected and in accordance with their role in the recovery phase. They included members of governmental institutions, civil society, and the market. The interview data were used for qualitative content analyses (Schnell et al. 2011; Flick 2012). In an email follow-up, interviewees could comment on the preliminary results and provide additional information on the current situation on-site. The data collected from the survey, the spatial analysis, and the interviews were triangulated in order to test their validity (Miles & Huberman 1994, p. 41). The data show in detail who was impacted, what damages were incurred to homes, who was insured, who received government assistance, and what the implications are.

Flood hazards, policy, and governance mechanisms

In Germany, floods are recurrent natural hazards with a significant economic impact. The majority of damage is due to riverine floods, both in the form of sudden flash floods or those impacting large river systems. This applies to the three major river catchment areas: the Rhine and Weser in the west, the Elbe and Oder in the east, and the Danube in the south. In the case of the latter, snowmelt or summer cyclone events contribute to riverine floods (Marx et al. 2017, pp. 9–10). These events are further exacerbated by precipitation on saturated soil or strong rainfall along sealed surfaces with low permeability, such as in urbanised areas with high degrees of coverage (Bronstert et al. 2017; pp. 87–99).

Flood hazards play an important role in German hazard legislation. The 1957 Federal Water Act (*Wasserhaushaltsgesetz*) and its notable revisions – following the 2002 floods and the European Floods Directive 2007/60/EC (Amtsblatt der Europäischen Union 2007; Government of the Federal Republic of Germany 2009) – indicate an increasing importance of flood adaptation for spatial planning. This reflects a shift of the concept of protection as a responsibility of the state towards managerial approaches based on the individual responsibility of mitigating risks to private property. However, state institutions seldom incentivise such individual responsibility in the context of preventing or managing potential flood disasters (Hartmann & Albrecht 2014, Thieken et al. 2016). Disaster management (*Katastrophenschutz*) spans the entire cycle of planning and preparedness, response, recovery, and reconstruction. In general, states receive support from the federal government in the case that a disaster is triggered by a natural and/or climate-related hazard. Since legislation lies with the states, frameworks, institutions, organisations, and measures can differ between them. This necessitates forms of cooperation between the federal and state levels of government (Isoard 2011; Vetter et al. 2017, p. 325).

From this point of view, the governance of flood hazards entails coordination and cooperation between various actors in order to provide information, assistance, and resources to impacted communities. The German strategy for civil protection (Bundesamt für Bevölkerungsschutz und Katastrophenhilfe 2010) emphasises preparedness and prevention and shared federal-state responsibilities during the response phase. Its guiding principle is subsidiarity, while relying heavily on the support of volunteer organisations (Marx et al. 2017, pp. 13–14). The governance of flood hazards in Germany takes place in the context of governmental crisis response (*Krisenbewältigung*). In the case of a disaster that was triggered by a natural hazard, response is a state responsibility and entails 'any civilian measure taken to protect the population and its livelihood from the impact of [...] disasters [...] as well as any measure taken to prevent, mitigate the impact of and cope with these events' (Marx et al. 2017, pp. 14–15).

Urban planning can influence the outcome of a flood disaster through related mitigation or adaptation measures. In particular, spatial and regional planning in Germany is required by law to meet sustainable development goals. This includes harmonising land uses, which is interpreted as a mandate for risk management tasks aimed at planning for and maintaining existing land uses in relation to spatially relevant risks. In this context, risk communication and participation are necessary to achieve a consensus on what constitutes an acceptable risk and which measures are suitable to reduce it. Informal planning procedures play an important role to enable participation, and, e.g., flood maps (as required by EU regulations) serve to communicate the related risks. Importantly, response, recovery, and reconstruction are not included within this scope, and the specification of land uses remains within the purview of development planning (*Bauleitplanung*) (Bundesinstitut für Bau -, Stadt- und Raumforschung 2015).

In Germany, climate change is a distinct policy field and the focus of a national strategy that acknowledges the influence of climate change on the increasing frequency and intensity of natural hazards such as floods (StMUV 2017; UNHSP 2017). The 2008 German Climate Change Adaptation Strategy and the 2011 Action Plan Adaptation illustrate this fact (Government of the Federal Republic of Germany 2008a; Government of the Federal Republic of Germany 2011). Nevertheless, adaptation to climate change and reduction of disaster risk still comprise separate structures, which leads to confusion and disconnects on municipal levels (Pauleit & Wamsler 2016, p.77). Due to increasingly intense and frequent flood hazards, certain land uses can become unsustainable, and relocation or resettlement (e.g. managed retreat) can become necessary. In this case, adaptation is supposed to take place before a disaster occurs or climate change impacts settlements and communities. Smaller communities may lack organisational capacities to achieve this, and compartmentalisation with regards to specific hazards comprises a further obstacle (Vetter et al. 2017, p. 325; Wamsler 2015, p. 30; Pauleit & Wamsler 2016).

Private flood insurance can be an important hazard governance mechanism both before and in the aftermath of a flood disaster, as it 'enables individual risk taking and provides a financial safety net in the face of uncertainty' (Krieger & Demeritt 2015, p. 4). It can support both disaster risk management and climate change adaptation approaches. Flood insurance in Germany is embedded within an all-hazards add-on insurance (*Elementarzusatzversicherung*) for private homeowners, who can purchase it voluntarily, yet coverage is relatively low (38 per cent of homeowners in 2015) (Krieger & Demeritt 2015, pp. 7–9; GDV 2015; Thieken et al. 2016). Households that purchase flood insurance are supposed to receive compensation in order to finance post-disaster rebuilding and reconstruction efforts (Ewald 1991, Krieger & Demeritt 2015, pp. 2–4; O'Malley 2004; Thieken et al. 2016). Only a few examples of flood insurance combined with risk mitigation exist or are incentivised by reduced insurance premiums or rewards (Thieken et al. 2016). Recurrent discussions on whether to create mandatory flood insurance remain unsuccessful on the grounds of the perception that the government would become the 'insurer of last resort' (Krieger & Demeritt 2015, p. 19). At the same time, it was decided that from 2019 onward governmental funding for rebuilding after disaster will no longer be available to homeowners if a related insurance product is available on the market (Kammerbauer & Wamsler 2018).

Summary of empirical research findings

Flood impacts on demography

The households that participated in the survey comprise 14 per cent of all households in the impacted area. 66 per cent of respondents are female, while the regional average is 51 per cent (Bayerisches Landesamt für Statistik und Datenverarbeitung 2013, p. 6). 96 per cent are citizens of the Federal Republic, matching the average in the governmental district of Lower Bavaria (Bayerisches Landesamt für Statistik und Datenverarbeitung 2014, p. 12). The remaining 4 per cent are from other EU countries. The average age of respondents is 54.5 years and is above the Lower Bavarian average of 44 years (Bayerisches Landesamt für Statistik 2016, p. 6). Our data indicate changes in employment, with a decrease from 28 per cent to 25 per cent. 15 participants are retirees at the time of the survey. Individual health is also a concern: 23 per cent state health is a significant problem and 34 per cent describe health as a very significant problem. 41 per cent report that health issues have started to occur after the flood.

Housing tenure

70 per cent of participants are long-term residents (20 years and longer). 76 per cent live in a single-family residential home. 81 per cent of survey

participants are homeowners before the flood. This share declines to 79 per cent after the flood. 9 per cent state that their homes are still uninhabitable at the time of the survey. 21 per cent own rental properties and some note a decrease of renters after the flood. The decrease in family names of multi-family residential buildings indicates that renters, who are typically dependent upon the decision of landlords to rebuild or not, left the area after the flood. Their motivation to return is different than in the case of homeowners who plan to rehabilitate the homes they own.

Flood damages

All respondents note damages to their homes, differentiated according to minor damage (4 per cent), major damage (70 per cent), and total destruction (26 per cent). 72 per cent were in the process of rehabilitating their homes at the time of the survey. 13 per cent were building a new home on the footprint of the previous (demolished) one. Two per cent were rebuilding on a different property.

Early recovery becomes impossible if homeowners have to demolish their homes, in some cases after rebuilding them, once a publicly appointed expert surveyor detects levels of oil contamination dangerous to human health. 72 per cent reported heating oil usage prior to the flood. 57 per cent of homes were contaminated with heating oil during the flood. At the time of the survey, 30 per cent of homes were still contaminated. 24 per cent were demolished. After the flood, 83 per cent of participants switched to other energy sources, and 15 per cent still used oil for heating. At the time of the survey, many respondents had not yet found a solution to the oil problem, pointing out a lack of financial support and consultation.

Recovery – with and without insurance

The possible influence that insurance has on rebuilding is limited to insured homeowners. While the majority of survey participants were homeowners, only 34 per cent of these had flood insurance, which is actually above the Lower Bavarian average of 30 per cent. The compound flood-oil contamination significantly exacerbated the impact of the disaster and the rebuilding needs that residents were confronted with. In some cases, insurance carriers dispute the degree of contamination. As a consequence of disputes, insurance money was paid late or in degrees that was insufficient to cover actual rebuilding costs. The degree of influence that insurance as a governance mechanism exerts on the recovery process is thus not only limited, but actually hampers the recovery efforts of insured homeowners. They see themselves at a disadvantage while their uninsured neighbours are able to rebuild with governmental funding.

66 per cent of households were informed about the financial support for uninsured homeowners. Applying for governmental funding was considered

difficult for 43 per cent and very difficult for 17 per cent. A majority of respondents were overwhelmed by the application processes that were perceived as too bureaucratic. The complex wording and requirements proved difficult to understand for many of the residents who had no higher education (47 possessed lower secondary education) or are migrants. Residents generally received help from volunteers when applying for assistance or other types of help. In Deggendorf, such help was particularly asked for by seniors, individuals with health concerns or disabilities, residents with a migration background, and renters who had been facing significant obstacles to their individual recovery. Beyond physical damage to structures and property, health and psychological trauma were further impediments to successful recovery and indicated the need for related assistance to support sustainable risk reduction and adaptation.

Discussion

Private insurance can be an important governance mechanism to manage the impacts of climate- and non-climate-related hazards, such as flooding. It is an inherent part of both disaster risk management and climate adaptation approaches as it has the potential to reduce impacts and vulnerability by enabling the prompt rehabilitation of damaged structures. In addition, it can promote the creation of resilient structures, both before and after a disaster occurs. However, since flood insurance coverage remains low in Germany, flooding repeatedly results in the need for state and federal governments to step in and provide funding for rebuilding houses of uninsured homeowners.

Our case study shows that vulnerability constitutes a major obstacle to recovery, due to the lack of coordinated action within the given governance configuration and the associated governmental and market actors, including insurance. The recovery in Deggendorf can be characterised as a return to normal or a reconstruction according to the status quo by 'rebuilding' physical and social structures that contributed to the disaster in the first place (Davis & Alexander 2016). The majority of homeowners decided to stay in the area. They have rehabilitated their homes, built them anew on the same site, or, to a lesser degree, have received replacement lots for rebuilding. The few homeowners who decided to change existing structures increased their resilience by raising structures or omitting residential uses on the ground floor. In addition, the installation of new oil heaters is now prohibited, but related oversight is limited. As a result, risks persist, and risk reduction among residents has been heterogeneous and ad-hoc (Olshansky & Chang 2009; Wamsler 2014; UNISDR 2015).

The recovery process in Deggendorf does not meet the requirements of 'building back better' (UNISDR 2015). Vulnerabilities continued to exist, and comprehensive and long-term risk reduction was not achieved. Variables of age, health, migration background, and access to financial resources or social networks were not addressed. They lead to challenges in the individual recovery of residents and had influenced their access to resources during and after the flood as well as their coping capacities (Wisner et al. 2004; Bohle 2001, Bohle & Glade 2008, p. 99). Importantly, the most pronounced differences in how quickly and successfully impacted residents were able to return and rebuild were rooted in whether they were homeowners or not and whether they were insured against floods or not.

Insured homeowners seemingly appeared as a contradiction in this case study. They received insurance payouts only late or after disputes, some of which resulted in settlements in court. In fact, our case study shows that insurance can disadvantage homeowners and slow down their recovery efforts (e.g. due to disagreements over issues, such as the severity of heating oil contamination). As a consequence, it can take citizens who have insurance longer to rebuild than uninsured homeowners who receive governmental assistance. They can be described as situationally vulnerable, (Bolin 2006, p. 125) which can be explained as an outcome of inadequate facilitation throughout all disaster management phases, where current governance systems enable settlement in at-risk locations (Kammerbauer & Wamsler 2018).

Conclusions

Based on our case study, the following observations can be made: The disconnect between different planning scales and the associated responsibilities and mandates of different actors to enable sustainable living environments did not support 'building back better' after disaster impact. Instead, it reproduced structural and social risk that had led to the disaster in the first place. This also relates to the fact that there is a disconnect between the constitutional obligation of the federal government to protect its citizens and the notion of individual responsibility promoted by EU water regulations and managerial approaches to disaster risk reduction. This translates into citizens' understanding that: 'if the state protects me, why do I need insurance?' Consequently, flood insurance coverage remains low. Moreover, our case study shows that uninsured homeowners are able to recover quicker than their insured counterparts. This is neither conducive to achieving consensus between the different actors of the current governance configuration, nor does it contribute to sustainable risk reduction and adaptation.

Our analysis leads to the following recommendations to support the role of flood insurance for effective flood risk governance and to foster sustainable and equitable risk reduction and adaptation: Market actors such as insurance carriers need to provide greater incentives for potential customers to increase coverage, reduce vulnerability and promote adaptation before and after the impacts of disaster. The governmental and civil society actors need to better coordinate with market actors, such as insurance carriers, to ensure synergy creation for improving disaster governance. Given that climate change will increase the number and intensity of floods in the future, the status quo simply is not enough, and actors of civil society, government, and the market need to communicate, coordinate, and cooperate with greater foresight in mind, by taking into consideration research outcomes as presented in this book chapter.

References

Amtsblatt der Europäischen Union 2007, 'Richtlinien', über die Bewertung und das Management von Hochwasserrisiken, <http://eur-lex.europa.eu/LexUriServ/LexUriServ.do?uri=OJ:L:2007:288:0027:0034:de:PDF>.

Bayerisches Landesamt für Statistik 2016, Regionalisierte Bevölkerungsvorausberechnung für Bayern bis 2035, Fürth, Beiträge zur Statistik Bayerns 548.

Bayerisches Landesamt für Statistik und Datenverarbeitung 2013, Eine Auswahl wichtiger statistischer Daten für den Landkreis Deggendorf, München.

Bayerisches Landesamt für Statistik und Datenverarbeitung 2014, Zensus 2011, Gemeindedaten Bevölkerung, Ergebnisse für Bayern, München.

Bohle, H.G 2001, 'Vulnerability and criticality: Perspectives from social geography', *IHDP Update*, vol. 2, no. 01, pp. 3–5.

Bohle, H.G, Glade, T 2008, 'Vulnerabilitätskonzepte in Sozial-und Naturwissenschaften', In Felgentreff, C, Glade, T (eds.), *Naturrisiken und Sozialkatastrophen*, Heidelberg, pp. 99–119.

Bolin, B 2006, 'Race, class, ethnicity, and disaster vulnerability', In *Handbook of disaster research*, Springer, New York, pp. 113–129.

Bronstert, A, Bormann, H, Bürger, G, Haberlandt, U, Hattermann, F, Heistermann, M, Petrow, T 2017, 'Hochwasser und Sturzfluten an Flüssen in Deutschland', In *Klimawandel in Deutschland*, Heidelberg, Germany, Springer Spektrum, pp. 87–101.

Bundesamt für Bevölkerungsschutz und Katastrophenhilfe 2010, *Pressemitteilung: Meilenstein in der Entwicklung des Bevölkerungsschutzes erreicht*, Bonn, Germany, Author.

Bundesinstitut für Bau-, Stadt- und Raumforschung 2015, Modellvorhaben der Raumordnung (MORO, Vorsorgendes Risikomanagement in der Regionalplanung. Endbericht. Berlin, Germany, pp. 25, 30.

Chambers, R 1989, 'Editorial introduction: Vulnerability, coping and policy', *IDS Bulletin*, vol. 20, no. 2, pp. 1–7.

Davis, I, Alexander, D 2016, *Recovery from disaster*, Abingdon, Oxon, Routledge.

Deutscher Bundestag 2013, Bericht zur Flutkatastrophe 2013: Katastrophenhilfe, Entschädigung, Wiederaufbau, Berlin, Drucksache 17/14743 vom 19, September 2013.

Ewald, F 1991, 'Insurance and risk', *The Foucault effect*, Chicago IL, University of Chicago Press; pp. 197–210.

Flick, U 2012, 'Qualitative Sozialforschung', In *Eine Einführung*, Reinbek bei Hamburg, Rowohlt.

Gerring, J 2007, *Case study research. Principles and practices*, New York, Cambridge University Press.

Gesamtverband der deutschen Versicherungswirtschaft, GDV 2015, 'Naturgefahrenreport 2015', Online-Serviceteil zum Berlin, Germany.

Government of the Federal Republic of Germany 2008a, 'Deutsche Anpassungsstrategie an den Klimawandel', Berlin, Germany, Deutsche Bundesregierung, <https://www.bmu.de/fileadmin/bmu-import/files/pdfs/allgemein/application/pdf/das_gesamt_bf.pdf>.

Government of the Federal Republic of Germany 2009, 'Gesetz zur Ordnung des Wasserhaushalts', In *Wasserhaushaltsgesetz*, WHG, Berlin, Germany, Bundesministerium der Justiz und für Verbraucherschutz, <http://www.gesetze-im-internet.de/bundesrecht/whg_2009/gesamt.pdf>.

Government of the Federal Republic of Germany 2011, 'Aktionsplan Anpassung der Deutschen Anpassungsstrategie an den Klimawandel', Berlin, Germany, Deutsche Bundesregierung, <http://www.bmub.bund.de/fileadmin/bmu-import/files/pdfs/allgemein/application/pdf/aktionsplan_anpassung_klimawandel_bf.pdf>.

Hartmann, T & Albrecht, J 2014, 'From flood protection to flood risk management: Condition-based and performance-based regulations in German water law', *Journal of Environmental Law*, vol. 26, no. 2, pp. 243–268.

Isoard, S 2011, 'Perspectives on adaptation to climate change in Europe', In Ford, J.D & Berrang-Ford, L (eds.), *Climate change adaptation in developed nations: From theory to practice*, Advances in Global Change Research 42. Heidelberg, Germany, Springer, pp. 51–68.

Kammerbauer, M 2019, 'Natural hazards governance in Germany', In *Oxford Research Encyclopedia*, Oxford, Oxford University Press.

Kammerbauer, M & Kaltenbach, F 2014, 'Sanierung bei Schadstoffbelastung in Überschwemmungsgebieten – Wann kommt das 'Hochwasserhaus'?' *Detail*, vol. 5, pp. 468–474.

Kammerbauer, M & Wamsler, C 2018, 'Risikomanagement ohne Risikominderung? Soziale Verwundbarkeit im Wiederaufbau nach Hochwasser in Deutschland', *Raumforschung und Raumordnung*, vol. 76, pp. 485–496.

Krieger, K & Demeritt, D 2015, 'Limits of insurance as risk governance: Market failures and disaster politics in German and British private flood insurance', Discussion Paper, 80. London, UK, Centre for Analysis of Risk and Regulation/London School of Economics and Political Science.

Landratsamt Deggendorf 2017a, 'Aktuelle Zahlen Aufbauhilfe Stand', 29 May 2017, Arbeitspapier des Landratsamts Deggendorf, Deggendorf.

Landratsamt Deggendorf 2017b, 'Zusammenfassung Hochwasser Deggendorf', Arbeitspapier des Landratsamts Deggendorf, Deggendorf.

Marx, S, Barbeito, G, Fleming, K, Petrovic, B, Pickl, S, Thieken, A & Zeidler, M 2017, Deutsches Komitee Katastrophenvorsorge, DKKV, 'Synthesis report on disaster risk reduction and climate change adaptation in Germany'.

Miles, M.B & Huberman, M.A 1994, *Qualitative data analysis. An expanded sourcebook*, Thousand Oaks, Sage, p. 41.

O'Malley, P 2004, *Risk, uncertainty and government*, London, UK, Routledge.

Olshansky, R & Chang, S 2009, 'Planning for disaster recovery: emerging research needs and challenges', *Progress in Planning*, vol. 72, no. 4, pp. 195–250.

Pauleit, S & Wamsler, C 2016, 'Making headway in climate policy mainstreaming and ecosystem-based adaptation: Two pioneering countries, different pathways, one goal', *Climatic Change*, vol. 137, pp. 71–87.

Schnell, R, Hill, P, Esser, E 2011, 'Methoden der empirischen Sozialforschung', München.

StMUV – Bayerisches Staatsministerium für Umwelt und Verbraucherschutz 2017, Integrale Konzepte zum kommunalen Sturzflut-Risikomanagement, München.

Susman, P, O'Keefe, P, Wisner, B 1983, 'Global disasters, a radical interpretation', In Hewitt, K (ed.), *Interpretations of calamity from the viewpoint of human ecology*, Boston, Routledge, pp. 263–283.

Thieken, A, Kienzler, S, Kreibich, H, Kuhlicke, C, Kunz, M, Mühr, B, Müller, M, Otto, A, Petrow, T, Pisi, S & Schröter, K 2016, 'Review of the flood risk management system in Germany after the major flood in 2013', *Ecology and Society*, vol. 21, no. 2, n.p.

UNDRR, *Vulnerability*, viewed 18 May 2018, <https://www.unisdr.org/we/inform/terminology>.

UNISDR 2015, 'Sendai Framework for Disaster Risk Reduction 2015–2030', <https://www.undrr.org/publication/sendai-framework-disaster-risk-reduction-2015-2030>.

United Nations Human Settlements Programme (UNHSP) 2017, 'Trends in urban resilience 2017', Nairobi, Africa, United Nations.

Vetter, A, Chrischilles, E, Eisenack, K, Kind, C, Mahrenholz, P & Pechan, A 2017, 'Anpassung an den Klimawandel als neues Politikfeld', In Brasseur, G.P, Jacob, D & Schuck-Zöller, S (eds.), *Klimawandel in Deutschland*, Entwicklung, Folgen, Risiken und Perspektiven, Berlin, Heidelberg, Germany, Springer Spektrum, pp. 325–334.

Wamsler, C 2014, *Cities, disaster risk and adaptation*, London, UK, Routledge.

Wamsler, C 2015, 'Mainstreaming ecosystem-based adaptation: Transformation toward sustainability in urban governance and planning', *Ecology and Society*, vol. 20, no. 2.

Wisner, B, Blaikie, P, Cannon, T & Davis, I 2004, *At risk: Natural hazards, people's vulnerability and disasters*, Abingdon, UK, Routledge.

Yin, R.K 2009, *Case study research. Design and methods*, vol. 5, Thousand Oaks, Sage.

Section III

Fire

Fire – our fourth element – is incredibly and obviously useful. Its affective visibility means that it is etched in memory and is ever-present – sparking the combustion engine, ringing the gas burner, stoking the campfire, and smouldering the cigarette. In heating, cooking, communing, forging, clearing, recreating, and regenerating, fire has been co-constitutive to human evolution and existence.

Fire is the flicking, flaring, travelling site of combustion and fuses together solid and gas in its being and becoming. In this dynamic, ofttimes unpredictable melding of air and earth, this element can be destructive. Its immense, hot-headed power conjures imaginings of purgatory and hell; of incessant burning and torment. Fire's ever-ready potential to bring forth terror and annihilation means that it manifests careful practices of domestication, suppression or protection within and around human settlements and in the home.

Fire insurance was one of the earliest formal insurances. It was pioneered in Britain following the 1666 Great Fire of London, financed by private funds – largely from merchants and manufacturers, and closely linked to the development and spread of marine insurance (Pearson 2012). In the nineteenth century British fire insurers lead the establishment of a global network of insurance and reinsurance (Pearson 2012) and in the United Kingdom, growing populations, increasing property values and new suburbs saw a steady rise in fire insurance premiums collected from 1 million in 1850 to 10 million by World War One. Taking out insurance to protect against losses caused by fire remains a steadfast element of the insurance portfolio.

Yet, despite its constitution through actuarialism, like other forms of insurance, contemporary fire-related products like house and contents insurance can appear far from certain. Booth and Harwood (2016, p. 50) describe how for residents in areas at high risk of wildfire, insurance is 'momentary rather than monetary.' Exceeding the financial, calculative logics of insurers, householder insurantial logics flicker in and out of focus, taking on different forms at different times and places within the complexities of everyday life (Booth 2021).

DOI: 10.4324/9781003157571-10

Insurance appears as a 'fire object'; manifesting as 'patterns of discontinuity between absence and presence' (Law & Singleton 2005, p. 331). It is other than what we might expect it to be, and as such can prove an elusive research subject (Booth 2021). A fire object,

> cannot be domesticated... faced with an object ... that is complex and generative in multiple and discontinuous absences, the limits of representational method are thrown into relief. We cannot bring it all to presence in conventional texts. We cannot bring it all to any particular presence. We cannot expect to be able to tell a consistent tale. And the implications of this? Other possibilities—for example the allegorical, the tolerance or art of ambiguity—might help. But in the first instance this suggests the need for methodological humility. If the world is messy we cannot know it by insisting that it is clear.
>
> (Law & Singleton 2005, pp. 349–350)

In this section, our authors consider insurance in the context of fire and the growing threat of wildfire, to lives and property in Australia and the United States. Collectively, these chapters illustrate the multifarious factors that constitute insurance and insurability, including the complex relations that exist between various agents and how this makes up insurance or contributes to underinsurance. Wildfire is less predictable than flood, and more destructive of property than wind or water. Photographs of razed homes from which all that remains are a few twisted sheets of corrugated metal, or a lone chimney left standing, are branded in the memories of those who live in landscapes of fire. Scott McKinnon, Christine Eriksen, and Eliza De Vet approach ideas of insurance and insurability from the standpoint of householders – specifically those who survived the 2013 Blue Mountain fires, and the 2020 Kangaroo Valley, Australia. They observe how the emotional as well as monetary valuing of household items contributes to a sense of insurability, and this draws into question ideas of underinsurance and 'adequate' insurance coverage as the value of insurance shifts and changes. Reflecting observations in previous chapters and those made by McKinnon and colleagues, Kenneth Klein highlights the role that insurers and insurance play in manifesting underinsurance and how householders and lending practices intersect with these, both in Australia and the United States. Again, in Australia, Pat O'Malley investigates the institutional setting of insuring against fire, observing how the idea of building sustainability is variously mobilized by developers, the fire protection industry and insurers, and illustrates the politically charged nature of insurance and insuring.

Within the fiery landscapes of Australia and the United States, there is a strong sense of insurance as a 'fire object' (Law & Singleton 2005). It can be described and illustrated in particular ways, yet there appears to be no consistent story to be told as agents – householders, governments, insurers – flicker in and out of focus with varying degrees of agency. In this context, insurance presents seemingly endless research opportunities – but not necessarily ones that permit neat answers to problematic phenomena such as underinsurance. Although our authors effectively illustrate key dimensions of insurance and insurability, the revealed complexity signposts a messiness and a boundlessness that can make insurance and underinsurance elusive research subjects.

References

Booth, K 2021, 'Critical insurance studies: Some geographic directions', *Progress in Human Geography*, vol. 45, no. 5, pp. 1295–1310.

Booth, K & Harwood, A 2016, 'Insurance as catastrophe: A geography of house and contents insurance in a bushfire prone area', *Geoforum*, vol. 69, pp. 44–52.

Law, J & Singleton, V 2005, 'Object lessons', *Organisation*, vol. 12, no. 3, pp. 331–355.

Pearson, R 2012, 'United Kingdom: Pioneering insurance internationally', In: Borscheid, P & Haueter, N.V (eds.) *World insurance: The evolution of a global risk network*, Oxford: Oxford University Press, pp. 67–97.

8 Between absence and presence

Questioning the value of insurance for bushfire recovery

Scott McKinnon, Christine Eriksen, and Eliza de Vet

Introduction

As the 'Black Summer' bushfires of late 2019 and early 2020 made clear, fire is an endemic element of the Australian landscape with often devastating, and now escalating, impacts on Australian households (Hughes et al. 2020). Insurance has become a critical tool in preparing for the threat of fire and responding to its impacts, partly because emergency management policies place significant responsibility on households to manage their own disaster risk (Eriksen et al. 2020). As a neoliberal mechanism for financialising risk, insurance has significant impacts on disaster policy and on disaster-impacted individuals (Collier 2014). Insurance can result in reduced opportunities for government, community, and household-led disaster mitigation (de Vet et al. 2019), while increasing economic inequality among disaster-impacted populations (Booth & Tranter 2018). In Australia, a significant proportion of the population is known to be underinsured, meaning that their insurance levels are unlikely to cover the substantial costs of rebuilding post-disaster (de Vet & Eriksen 2020; Booth & Tranter 2018). Nonetheless, insurance contributes to the resilience of many bushfire survivors by increasing their financial capacity to rebuild (Eriksen & de Vet 2020). It thereby mitigates the substantial stress and anxiety of post-disaster rebuilding processes (Dixon et al. 2015). Insurance thus plays a valuable role in recovery, and yet that value is often ambiguous and contingent on a range of non-financial factors.

In this chapter, we attend to bushfire survivor accounts at two case study sites in order to explore the uncertain role of insurance within recovery processes. In particular, we consider how bushfire survivors assess and define both the value of insurance and the insurability of valued objects. Survivors often prioritise intangible and uninsurable currencies – including wellbeing, time, and labour – as being of more value than the material possessions or built structures covered by insurance policies. Equally, many survivors describe a changed relationship to the material and insurable objects that have been made literally absent (yet which remain emotionally or mnemonically present) by the flames (Eriksen & Ballard 2020; McKinnon

DOI: 10.4324/9781003157571-11

& Eriksen 2021; see also Booth 2020). As a result, survivors often assess the value of destroyed or damaged objects in terms that are incompatible with financialised models of insurable risk and loss.

In part, the disjuncture between value as defined by an insurer and by a disaster survivor relates to discordant definitions of home. Geographer Alison Blunt (2005, p. 506) defines home as 'a material and an affective space, shaped by everyday practices, lived experiences, social relations, memories and emotions.' While insurance may support the rebuilding of material structures, its capacity to mitigate the impacts of disaster on the affective, mnemonic, and emotional aspects of home is limited (Eriksen & de Vet 2020). A home has the potential to act as a site of belonging (Gorman-Murray 2011), and the loss of home in disaster has troubling potential consequences on mental health and emotional wellbeing (Camilleri et al. 2010). Faced with the destruction of their home, bushfire survivors are unlikely to view their loss as a purely financial one. In the days, weeks, and months after a disaster, the assessment of monetary values by insurers, and the emotional impacts of loss on householders, are an uneasy combination with significant impacts on survivor wellbeing.

Bushfire-related insurance in Australia generally delineates between 'home' insurance, which relates to the material structures of a house, and 'contents' insurance, which relates to the possessions stored within the house's physical structure. This distinction between a building and the objects within it contrasts with the ways in which a house and material possessions (along with the landscape where they are located) are intertwined as critical elements in understandings of home as an affective space and site of belonging (Blunt & Dowling 2006; Booth 2020; Hurdley 2013). Home is created, in part, through 'emotional encounters with domestic objects, interiors and furnishings' (Walker 2020, p. 13). Material objects displayed in homes act as mnemonic anchors, creating a sense of shared identity for members of the household by drawing on memories of shared interests or experiences (de Vet, Eriksen & McKinnon 2021). Similarly, household items connect to identities at broader scales, including to minority identities or diasporic communities (McKinnon, Gorman-Murray & Dominey-Howes 2016; Tolia-Kelly 2004). Carrying cherished possessions between homes is an important element of homemaking and allows for continued connections to the past (Pazhoothundathil & Bailey 2020). A fear for many disaster survivors is that the loss of belongings will destabilise the memories and identities to which they are attached. This will, in turn, trouble feelings of safety and security at home (Gorman-Murray, McKinnon & Dominey-Howes 2014; McKinnon & Eriksen 2021). The consequences of this loss sit outside insurable values.

According to Dowling and Mee (2007, p. 161), 'Home is as much a process as it is a thing.' The process of homemaking, which includes both a home's creation and its maintenance, requires substantial investments of time and labour. The ruin of a bushfire-destroyed home represents the loss of those past investments, along with a significant demand for future time in recreating what was lost. Time is therefore highly valuable to bushfire survivors. The ways in which insurance may or may not enable the purchase of time and labour or, conversely, may cost more time through complex application processes and disputed negotiations (Whittle et al. 2012), are critical to survivor assessments of its value.

Attending to survivor accounts allows for a deeper understanding of the entwining of home, recovery, and insurance, and suggests the need to understand both the value and the cost of insurance in more than monetary terms. In making assessments both of the value of insurance, and of the insurability of belongings, disaster survivors do not always position or configure insurance as a purely financial mechanism. Rather, the value or cost of insurance is assessed in more than monetary terms, in which negative impacts on wellbeing, the loss of time, and the nonfinancial value of belongings, each often take priority. In other words, amid the emotional complexity of recovery processes, the value of insurance is what people make it.

Below, we outline our case study sites and methods. In order to explore questions of insurance and 'value' in more than monetary terms, we have then divided our analysis into three sections. First, we explore impacts of insurance on survivor wellbeing. We consider how bushfire survivors assessed the value of insurance based both on the emotional and the financial support they received. Second, we consider the insurability of material belongings within the home, and explore tensions between the value of an object as assessed by an insurer and a bushfire survivor respectively. Finally, we examine how bushfire survivors prioritised intangible and uninsurable objects, such as time and labour, in ways that challenge ideas of insurable value. In each case, we highlight the agency of bushfire survivors in actively assessing the value of insurance, particularly as it relates to how bushfire and insurance affected the absence and presence of home – materially and emotionally.

Case studies and methods

We draw on qualitative analysis of in-depth interviews with bushfire survivors in two case study sites in New South Wales, Australia. The Blue Mountains and Kangaroo Valley are both fire-prone areas, and both have a history of damaging bushfires. In October 2013, bushfires in the Blue Mountains, on the western outskirts of the Greater Sydney region, destroyed 203 houses and damaged a further 286. Fourteen semi-structured interviews with 17 Blue Mountains residents (including three couples) were conducted four years after the fires, in late 2017 and early 2018. During the summer of 2019-2020, bushfires of unprecedented scale burnt across massive areas of south-east Australia, destroying 2,439 homes in New South Wales and resulting in the death of 33 people and a billion animals (Hughes et al. 2020). This included parts of the Blue Mountains as well as the Shoalhaven region to the south, where Kangaroo Valley is nestled into the foot of the same mountain range. On 4 January 2020, the Currowan fire, which had burnt across 320,845 ha in the previous 74 days, tore through the western side of Kangaroo Valley. In mid-2020, 14 interviews were conducted with 19 Kangaroo Valley residents (including five couples).

Across the case study sites, participants' homes were located on a range of property types, including farms, holiday accommodation businesses, and blocks of varying sizes in rural, suburban, and peri-urban areas. Experiences of the fires also varied. Some participants stayed and defended their homes through the fires, often in terrifying circumstances. Others evacuated before the fires and returned home in subsequent days to find scorched landscapes and, in several cases, destroyed homes. All participants had some form of insurance policy or policies, be it home, contents, and/or farm insurance. In both case studies, interviews were recorded and transcribed verbatim. Using the QSR NVivo v.11 qualitative data analysis software, transcripts were coded and thematically analysed. The names of participants included in this chapter are all pseudonyms.

Insurance and wellbeing

Insurance affected the wellbeing of participants in multiple and complex ways, which were often disconnected from the monetary value of a payout. An important contributing factor was how much value insurers were perceived to place on the mental health and emotional wellbeing of their clients (Eriksen & de Vet 2020). In other words, did insurance assessors appear to act purely as an extension of a financial mechanism, or did they actively and visibly support the emotional needs of potentially traumatised individuals? David and Neal (Kangaroo Valley (KV), 50s), for example, lost their home and a holiday rental in the Currowan fire. They were appreciative of the speed with which their claim was processed by their insurer, which had allowed them to begin plans for rebuilding relatively quickly. Also of significance in their assessment of the value of insurance, was the empathy and consideration of the insurance assessors who worked with them. David praised 'the one-on-one genuine compassion that was shown,' while Neal stated, 'To get hugs from assessors who are concerned about your welfare at a time of trauma was quite lovely.'

Similarly, when Jenny and Frank (KV, 50s) were asked about the process of claiming insurance on their destroyed home, their immediate response was to highlight the emotional support they received from their assessor. Jenny stated, 'I mean, I've got to know the assessor woman so well she's coming down for coffee [laughing] … She's like a sister, you know, she's just, we ring up and they were so kind and it was so humbling.' The couple's insurance payout was ultimately not enough to cover the costs of their planned rebuild. However, they attributed this problem to the large development approval fees from the local government, rather than limitations in the insurance itself. Ultimately, Jenny and Frank experienced insurance as a form of emotional, as well as financial support. Their insurance payout was certainly significant to their ability to rebuild and recover, but it was far from the only criteria through which they assessed the value of insurance.

This more than monetary assessment of insurance was equally highlighted by the negative experiences of some participants, who found that difficult and even hostile interactions with insurance providers carried significant emotional costs. Kenneth (Blue Mountains (BMs), 50s) lost his onsite holiday accommodation in the 2013 bushfires. His subsequent attempts to claim on insurance became extremely complicated and his relationship to assessors – who he described as 'bossy' and 'arrogant' – was more antagonistic than supportive. Similarly, Margaret (KV, 60s) was involved in a difficult series of disputes with her insurers and stated, 'But the insurance company's being an insurance company, they like to, you know, make you suffer.' This suggests an understanding of insurance as a deliberately hostile or antagonistic process, the potential financial benefit of which came at a significant emotional cost.

The above examples reveal differing ways in which insurance became present in the lives of the bushfire survivors. For some, it was a reassuring presence, which made itself felt through acts of concern and kindness. In such circumstances, insurance was valued as a counterbalance to the negative impacts of the bushfire – a beneficial support towards a post-fire life. For other participants, insurance was entwined with the multiple negative, stressful, and anxiety-producing objects produced by the bushfire. Rather than being an element of disaster recovery, it was an enduring element of the disaster itself, and one more obstacle to negotiate in the struggle for a post-fire life.

Absence, presence, and material belongings

A core promise of contents insurance is that destroyed material belongings are never entirely lost, but can instead be replaced post-fire. Within fire-prone landscapes, insurance thus provides at least some reassurance that objects will be 'safe,' while simultaneously positioning uninsured objects as 'unsafe' (Booth 2020). This promise rests on three assumptions – that destroyed belongings are: i) insurable to a value at which an equally valued replacement can be found; ii) will be considered worthy of replacement by survivors post-fire; and iii) are of purely monetary value to their owner. However, in both case studies survivors described relationships to lost belongings that complicated, or even subverted, these assumptions. Survivors described changed relationships to material belongings, while at times rejecting the idea that particular objects were replaceable at all – materially and emotionally.

Several participants described how the bushfire had prompted them to reconsider their need or desire to re-fill their homes with material objects in the future. In these cases, domestic belongings were reframed less as fragile possessions needing protection from fire, and instead as heavy constraints or burdens from which survivors had now been freed (see also Horton & Kraftl 2012; McKinnon & Eriksen 2021). Ronald (BM, 70s) used dark humour to describe this transformed relationship, labelling the disaster's destruction a 'fire sale.' Rather than damaging, he preferred to think of the fire as cleansing, in that it had 'cleaned all the clutter' from his home. The idea that material belongings were 'clutter' was repeated by Heather (BM, 40s) who stated, 'I think you learn, you declutter and you simplify your life if that sort of thing happens.'

This re-framing by survivors brings in to question the assessment of insurance levels as either adequate or inadequate. A survivor without sufficient insurance to replace lost home contents might be labelled 'underinsured.' Yet, if they have no intention to replace all their belongings and, indeed, feel freed from, rather than lost without, these belongings, then the term 'underinsurance' has little relevance to their lived experience with fire.

This is not to suggest that Heather and other participants who expressed similar sentiments had not experienced sadness or even trauma at their loss. The reassessment of objects as 'clutter' was often part of the recovery process, moving from mourning for enormous losses towards an acceptance of transformed meanings and relationships. Bill and Sue (KV, 50s), for example, also lost their house and its contents. Asked about lost belongings, Sue replied, 'Now we just call it stuff, we don't think about it too much, but it was heartbreaking.' Eriksen and Ballard (2020, p. 78) argue that post-fire, a material object 'is often understood through its absence.' In Sue's description, material belongings have a ghostly presence, which is not replaceable but rather called upon post-fire for remembered non-monetary values and for the emotional costs of loss. Subsequently re-labelled 'clutter' and 'stuff,' these objects are dismissed as unknown and unnamed, and can therefore be more easily dealt with as absent.

Sue recalled that many of their lost belongings had been covered by insurance, before stating, 'not that [insurance] covers the sentimental value.' This statement reveals how objects insurance companies consider insurable, may be considered uninsurable by bushfire survivors because of their non-monetary or more than monetary value (Booth & Harwood 2016). A similar sentiment was expressed by Helen (KV, 60s), who had stored a number of precious belongings in a shelter below her home, as the Currowan fire approached. She and her husband, Joe, had insured the buildings on their property but not the contents. They stayed and defended the property through the fire and their home suffered little damage. When asked if they now intended to change their insurance levels, Joe stated, 'If anything, I'd try and reduce it. We've had the fire now. That was really the only risk that I thought we had.' Helen added that, while insurance could help rebuild a building, 'It couldn't replace all my treasures that I keep looking at saying, "Oh thank God that's still there. I forgot to evacuate that."'

To Helen, contents insurance was unnecessary because her 'treasures' were irreplaceable and could not be assigned a monetary value. From this point of view, no level of insurance could be considered adequate. Although the fire caused Helen to imagine the possible absence of treasured belongings, and to feel frequent relief about their continued presence as elements of her home, she nonetheless rejected the idea that insurance would have made them in any sense safer. For Helen and Joe, the value of insurance was only in its ability to 'build another building.' It offered no value in protecting the belongings that made the building feel like home.

Part of the process of obtaining contents insurance is a stocktake of household belongings, and an assessment of their monetary value should they be destroyed and need to be replaced. The accounts of bushfire survivors complicate this seemingly logical and linear financialised process – both by questioning whether there is value in replacing lost things, and by rejecting the idea that particular belongings can be assigned a monetary value. By prioritising emotional values over financial costs, participants in both case studies challenge ideas of insurable value, particularly as it relates to how bushfire and insurance variously affect the absence and presence of home.

Intangible objects

While home takes material form through objects and built structures, equally significant to the making of home are the time, skill, and labour that go into its construction and maintenance. Homemaking incorporates a range of (often highly gendered) activities including daily routines, DIY and other forms of construction, gardening, decorating, and cleaning (Blunt & Dowling 2006; Baxter & Brickell 2014). Attentiveness to these acts reveals the home as an ongoing process of everyday practices (Blunt 2005). In both case studies, participants highlighted 'the emotional intensity of the post-disaster environment' (Morrice 2013, p. 34), and contemplated how intangible and non-financial objects, such as the time and labour involved in processes of homemaking, sat outside – and even conflicted with – insurable values.

Concern for the value of intangible objects, such as time, was also linked to the emotional costs of losing a home. Insurance might financially compensate for the time of architects, builders and others hired to rebuild a burnt-out home. Yet, insurance cannot replace or compensate for the years of effort invested by householders in creating, maintaining and, in some cases, literally building their now lost home. While mourning this lost time and effort, participants also had to come to terms with a future where more time and labour would be required simply to re-establish what they had already achieved before the fire struck.

Tony (KV, 50s), for example, stayed and defended his home from the Currowan fire – a traumatic experience that had affected his mental health. Although the house was saved, the extensive native gardens he had cultivated, which blurred the visual boundary between his property and surrounding bushland, were destroyed. Insurance did not cover the garden, and the many hundreds of burnt plants would be costly to replace. Just as significant, in Tony's estimation, were the time and labour involved in establishing the gardens over several years. He stated, 'What I've said to people is that if the house had been damaged, we could put an insurance claim in and rebuild it ... You can't replace the garden because I've just lost that time and that effort and that money and that's a start all over again thing.' Gardening can be seen here as a form of home-making practice (Reid & Beilin 2015), and in Tony's description, the garden and the home are indistinguishable. Yet in insurantial terms, they are entirely separate. Tony had lost the beauty of surrounding trees and shrubs, which were now an ashen and charred environment. Also lost was the substantial effort of establishing and maintaining the garden over many years. Insurance covered none of the entwined physical, emotional, and financial costs of this loss.

Michael (BM, 30s) was living in rental property at the time of the fires. A shed on the property was destroyed, but the house survived. He had subsequently built a home and, in discussing plans for insurance, Michael positioned his own labour as a complicating factor in assessing insurable value. A skilled builder, Michael had worked on the design and construction of the home, which he saw as part of an emotional connection both to the site at which the property was located, and to the process of homemaking. He contrasted this with the choice to build a so-called 'McMansion' (a pejorative term for a large, mass-produced house), which could just be 'plonked' on a site by a building company – a process Michael saw as 'all very clinical. There's no emotion, there's no connection.' Unfortunately, his preferred labour-intensive and emotionally rich process resulted in a cherished dwelling that was not easily insurable. Through his own labour, Michael had been able to build a home that, in monetary terms, was more valuable than he would have been able to afford otherwise. As a result, in Michael's words, 'the bloody premiums are going to be nuts, just nuts,' and were beyond his financial means. Because of these complicating factors, Michael was considering being 'a little bit fatalistic,' and insuring the home for less than its replacement value. If he lost the home in a future fire, he would again invest his own labour, and cover the financial shortfall from any payout by building back smaller next time. These considerations again reveal the complex meaning of value, as the physical and emotional process of homemaking often does not fit, or plainly conflicts with, insurantial values.

It is important to note how wellbeing, material belongings, and intangible objects are also deeply connected in survivor assessments of the value of insurance. Ben (KV, 50s) had almost completed a 12-year process of building his own home when the Currowan fires struck. Fortunately, the near-complete home survived but a storage shed on the property, along with a small cabin in which Ben had been living, were both destroyed. Ben had paid little attention to insurance levels over the preceding years. He was pleasantly surprised to discover that his lost belongings were insured to a higher level than he had expected. Although the loss of his cabin and its contents, as well as the devastating impacts on the surrounding bushland, had been painful experiences for Ben, the unexpectedly large financial payout, and the ease of the process of filling in his claim, had provided him with valuable time with which to recover. Rather than replacing lost items, Ben instead used his insurance payout to finance time off work, during which he cleared the fire-damaged parts of his property, and worked on finishing his home. These acts of post-fire homemaking and engagement with the transformed property were important to Ben's recovery process, and were of more significant value to him than replacing lost belongings.

Attending to these bushfire survivor accounts suggests the need to understand time and labour, particularly as they relate to processes of homemaking, as intangible objects of enormous value that nonetheless remain uninsurable. A significant impact of bushfire is lost time, both in terms of the lost time that went into making the home pre-fire, and the time it will take to make a post-fire life. In some cases, insurance reduced this impact, including through payouts that funded the purchase of labour and time from builders. In others, insurance took up more time, and required more effort in navigating complex policies and negotiating disputes with assessors. Again, the bushfire survivors calculated the value of insurance to their recovery via factors that were either non-monetary or that circumvented financial hurdles.

Conclusion

As argued by Whittle et al. (2012, p. 68), 'recovery is more than just a 'bricks and mortar' exercise to be measured through statistics of displacement and return or economic damage.' Equally, a home is more than simply a set of material (and insurable) structures. It is also a spatial and temporal process of creation and maintenance, enfolded with emotional, mnemonic, and affective meanings. The material destruction of a home by fire is a devastating loss, and witnessing or engaging with the ruins is often a confronting yet important moment for survivors (Schlunke 2016; Eriksen & Ballard 2020). For insured householders, insurance offers at least some confidence in their financial capacity to rebuild. Lost material structures and belongings can be imagined into presence in a rebuilt and restored future. Yet, impacts on emotional wellbeing, intangible currencies, and irreplaceable objects must equally be negotiated through the recovery process. In this chapter, we have explored how survivors negotiate the disjuncture between their understanding of home and loss, and the ways in which that loss is configured by insurance only as a 'bricks and mortar' exercise to be assessed, compensated, and rebuilt.

In neoliberal models of addressing disaster risk through market-based financialisation, the impacts of bushfire on the non-monetary meanings and values of home are absent. Insurance is known to exacerbate social inequalities pre- and post-disaster (Booth & Tranter 2018), with consequent impacts on mental health and escalating challenges to recovery processes (de Vet & Eriksen 2020; Eriksen & de Vet 2020). Opaque insurance policies and non-negotiable policy terms equally result in inequitable power dynamics between insurance companies and policyholders, often resulting in high levels of distrust and uncertainty (Booth & Harwood 2016). Nonetheless, the stories of bushfire survivors in New South Wales offer insights into the agency of insured householders, and how uneven power structures, and discordant understandings of home, are negotiated and reconstructed by survivors post-fire. This is not to say that the actions of insurers do not, at times, negatively impact recovery processes, or that insurance payouts are unimportant to the ability to rebuild and recover. Rather, many bushfire survivors draw emotional support from insurance providers, while also prioritising the non-monetary value of time and labour, along with the emotional, mnemonic, and affective value of material objects.

The impacts of climate change on the frequency and intensity of bushfires, along with growing populations in fire-prone landscapes, mean more Australian households than ever before are at risk. Given Australia's fiery future, it is crucial to reconsider the neoliberal financialising of risk, so it can more adequately address the needs of future disaster survivors. What the bushfire survivor accounts examined in this chapter ultimately suggest, is that disaster risk is more than monetary, and that the destruction of material structures and objects has impacts that resonate far beyond the monetary cost of rebuilding and replacing a house and its contents. Developing and adequately resourcing support systems, which in addition to addressing practical needs also meet the psychosocial needs of survivors, is essential (Eriksen and de Vet 2020). An important step towards understanding these needs is to comprehend both the value and cost of insurance to recovery processes in more than monetary terms.

References

Baxter, R & Brickell, K 2014, 'For home unmaking', *Home Cultures*, vol. 11, no. 2, pp. 133–143.

Blunt, Alison 2005, 'Cultural geography: Cultural geographies of home', *Progress in Human Geography*, vol. 29, no. 4, pp. 505–515.

Blunt, A & Dowling, R 2006, *Home*. New York, NY: Routledge.

Booth, K 2020, 'Firescapes of disruption: An absence of insurance in landscapes of fire', *Environment and Planning E: Nature and Space*, vol. 4, no. 2, pp. 525–544.

Booth, K & Harwood, A 2016, 'Insurance as catastrophe: A geography of house and contents insurance in bushfire-prone places', *Geoforum*, vol. 69, pp. 4–52.

Booth, K & Tranter, B 2018, 'When disaster strikes: Under-insurance in Australian households', *Urban Studies*, vol. 55, no. 14, pp. 3135–3150.

Camilleri, P, Healy, C, Macdonald, E.M, Nicholls, S, Sykes, J, Winkworth, G & Woodward, M 2010, 'Recovery from bushfires: The experience of the 2003 Canberra bushfires three years after', *Journal of Emergency Primary Health Care*, vol. 8, no. 1, pp. 990383-1–990383-15.

Collier, S.J 2014, 'Neoliberalism and natural disaster', *Journal of Cultural Economy*, vol. 7, no. 3, pp. 273–290.

de Vet, E & Eriksen, C 2020, 'When insurance and goodwill are not enough: Bushfire Attack Level (BAL) ratings, risk calculations and disaster resilience in Australia', *Australian Geographer*, vol. 51, no. 1, pp. 35–51.

de Vet, E, Eriksen, C, Booth, K & French, S 2019, 'An unmitigated disaster: Shifting from response and recovery to mitigation for an insurable future', *International Journal of Disaster Risk Science*, vol. 10, pp. 179–92.

de Vet, E, Eriksen, C & McKinnon, S 2021, 'Dilemmas, decision-making and disasters: Emotions of parenting, safety and rebuilding in bushfire recovery', *Area*, vol. 53, no. 2, pp. 283–291.

Dixon, K.M, Shochet, I.M & Shakespeare-Finch, J 2015, 'Stress during the rebuilding phase influenced mental health following two Queensland flood disasters more than the event itself', In: *Australian and New Zealand Disaster and Emergency Management Conference*, Gold Coast, Australia, 3–5 May.

Dowling, R & Mee, K 2007, 'Home and homemaking in contemporary Australia', *Housing, Theory and Society*, vol. 24, no. 3, pp. 161–165.

Eriksen, C & Ballard, S 2020, *Alliances in the Anthropocene: Fire, plant and people*, Singapore: Palgrave Macmillan.

Eriksen, C & de Vet, E 2020, 'Untangling insurance, rebuilding and wellbeing in bushfire recovery', *Geographical Research*, vol. 59, no. 2, pp. 228–241.

Eriksen, C, McKinnon, S & de Vet, E 2020, 'Why insurance matters in emergency management: insights from disaster research', *Australian Journal of Emergency Management*, vol. 35, no. 4, pp. 42–47.

Gorman-Murray, A 2011, 'Economic crises and emotional fallout: Work, home and men's senses of belonging in post-GFC Sydney', *Emotion, Space and Society*, vol. 4, no. 4, pp. 211–220.

Gorman-Murray, A, McKinnon, S & Dominey-Howes, D 2014, 'Queer domicide: LGBT displacement and home loss in natural disaster impact, response, and recovery', *Home Cultures*, vol. 11, no. 2, pp. 237–261.

Horton, J & Kraftl, P 2012, 'Clearing out a cupboard: Memory, materiality and transitions', In Jones, O. & Garden-Hansen, J (eds.), *Geography and memory*, London, UK: Palgrave Macmillan, pp. 25–44.

Hughes, L, Steffen, W, Mullins, G, Dean, A, Weisbrot, E & Rice, M 2020, *Summer of Crisis*, Climate Council of Australia Limited, At: https://www.climatecouncil.org.au/wp-content/uploads/2020/03/Crisis-Summer-Report-200311.pdf.

Hurdley, R 2013, *Home, materiality, memory and belonging: Keeping culture*, Basingstoke, UK: Palgrave MacMillan.

McKinnon, S & Eriksen, C 2021, 'Engaging with the home-in-ruins: Memory, temporality and the unmaking of home after fire', *Social and Cultural Geography*, DOI: 10.1080/14649365.2021.1939127.

McKinnon, S, Gorman-Murray, A & Dominey-Howes, D 2016, '"The Greatest Loss was a Loss of our History": Natural disasters, marginalised identities and sites of memory', *Social and Cultural Geography*, vol. 17, no. 8, pp. 1120–1139.

Morrice, S.J 2013, 'Heartache and Hurricane Katrina: Recognising the influence of emotion in post-disaster return decisions', *Area*, vol. 45, no. 1, pp. 33–39.

Pazhoothundathil, N & Bailey, A 2020, 'Cherished possessions, home-making practices and aging in care homes in Kerala, India', *Emotion, Space and Society*, vol. 36, p. 100706.

Reid, K & Beilin, R 2015, 'Making the landscape 'home': Narratives of bushfire and place in Australia', *Geoforum*, vol. 58, pp. 95–103.

Schlunke, K 2016, 'Burnt houses and the haunted home: Reconfiguring the ruin in Australia', In Cook, N, Davison, A & Crabtree, L (eds.), *Housing and home unbound: Intersections in economics, environment and politics in Australia*, Oxford, UK: Routledge, pp. 218–231.

Tolia-Kelly, D 2004, 'Locating processes of identification: Studying the precipitates of re-memory through artefacts in the British Asian home', *Transactions of the Institute of British Geographers*, vol. 29, no. 3, pp. 314–329.

Walker, A 2020, "I don't know where all the cutlery is': Exploring materiality and homemaking in post-separation families', *Social & Cultural Geography*, DOI: 10.1080/14649365.2019.1705995.

Whittle, R, Walker, M, Medd, W & Mort, M 2012, 'Flood of emotions: Emotional work and long-term disaster recovery', *Emotion, Space and Society*, vol. 5, no. 1, pp. 60–69.

9 Is fire insurable?

Insights from bushfires in Australia and wildfires in the United States

Kenneth S. Klein

Fire and insurance have been conjoined for a very long time. On 2 September 1666, the Great Fire of London began. Estimates are that when the fire was done four days later, 70,000 of the City's 80,000 inhabitants were homeless. And at least myth – perhaps reality – has it that in the immediate next several years, out of the ashes of that fire the idea of the first fire insurance company germinated to fruition in the mind of Nicholas If-Jesus-Christ-had-not-died-for-thee-thou-hadst-been-damned Barbon (James 1954, pp. 44–45). In the following 450+ years, fire and insurance have taken a journey together in an inter-relationship that continues to evolve.

Today, most homeowners want full and adequate fire insurance, are willing to pay for it, think they have it, and yet do not. Whether 'bush-fire' in Australia or 'wildfire' in the United States, the frequency, intensity, and economic impacts of catastrophic fire events are increasing. The State of California, for example, now essentially has a year-round fire season (CalFire 2021). And the consequence is that dwelling insurance is becoming less affordable, less available, and less adequate. Ubiquitously affordable, adequate, available dwelling insurance is an aspiration that seems more remote now than ever, and yet also more necessary than ever.

Homeowners want to and think they have fully insured their dwellings for fire

Homeowners want to fully insure their homes, and until disaster strikes, think they have done so. It is postulated that one reason, 'individuals do not buy insurance is that they perceive the probability of a loss to be below their threshold level of concern so that the benefits of insurance exceed the associated premium and search costs' (Kunreuther 2018, p. 143). Depending upon the theorist, this sometimes may be described as an adverse selection problem, or price elasticity. Taken out of economics jargon, it is theorising that one reason there may be uninsureds or underinsureds is that individuals do not want to share the cost of someone else's risk.

The theory is intuitively plausible, but apparently at least for homeowners deciding about insuring their dwellings for fire, the theory is wrong.

DOI: 10.4324/9781003157571-12

Amongst homeowners who have dwelling insurance for fire, most want adequate (meaning, full) insurance. And most homeowners who have a choice, choose to insure their dwelling for fire.

Simply put, homeowners typically do not choose to underinsure their dwelling. Most homeowners want to fully insure or over-insure. This is an incidental but important finding of work by economists Benjamin Collier and Marc Ragin (Collier & Ragin 2019, Table 9.3). They studied the National Flood Insurance Program (NFIP) in the United States. It is a public insurance product, sold by private insurance agents. In other words, all insurers offer the same product – the only variable is the agent selling it. Collier and Ragin were interested in the NFIP for this reason ¬ they were trying to study the influence of an insurance agent on the decision of how much insurance to buy. And the NFIP Program lets them control for all other variables. The NFIP offers maximum cover of $250,000, and a minimum cover of 80% of the estimated rebuild cost or of $250,000, whichever is less. Collier and Ragin isolated policies with estimated rebuild costs of less than $250,000. In other words, in these instances, insureds had a choice of 20% underinsuring, insuring to estimated rebuild cost, or over-insuring up to $250,000. 80% of homeowners either insured to the insurer's estimated rebuild costs, or over-insured above that. And in Australia a survey of homeowners affected by the ACT bushfires found an identical number – 80% said they were adequately insured (Australian Securities & Investment Commission (ASIC) 2005, p. 63).

And most homeowners – either voluntarily or involuntarily – do insure their homes for fire. 'Homeowner' or 'householders' insurance, as the product denomination implies, provides cover for the owner of a dwelling in the instance of damage or destruction of the dwelling (Federal Insurance Office (FIO) 2015, pp. 13, 15–20; Australian Competition & Consumer Commission (ACCC) 2020, pp. 12–17). Homeowner insurance may insure both the dwelling and/or the contents of the dwelling. For homes with a mortgage, however, insurance of the structure – the collateral for a mortgage loan – is not a choice; mortgages in both Australia and the United States require the homeowner have insurance of the mortgaged dwelling for fire (ACCC 2020, p. 147; FIO 2015, pp. 3, 15).

Because of the architecture of mortgages in the United States, there almost *always* is fire insurance of the dwelling in place for homes with a mortgage. Most mortgages in the United States provide that if the borrower allows insurance to lapse then the lender will purchase 'force-placed' insurance at the borrower's expense; this insurance protects the lender from a fire loss of collateral (Cronkite 2016, p. 691). There does not appear to be an analogue in Australia to force-placed insurance, where it seems at least theoretically possible for a home under mortgage to have no insurance of the dwelling for fire (ACCC 2020, p. 462).

Nonetheless, the prevalence of insurance of dwellings for fire is exceptionally high and nearly identical in both the United States and Australia. In the

United States, over 90% of homes – perhaps as high as 95% – have home-owner insurance (Insurance Information Institute (III) 2016). In Australia, 89%–96% of homeowners have an insured property (Booth & Tranter 2018, p. 3137; ACCC 2020, p. 269). In other words, the presence or absence of force-placed insurance mechanisms does not seem to impact the likelihood of whether a mortgaged dwelling is insured. In both Australia and the United States, all or almost all mortgaged homes are insured for fire if for no other reason than they have to be.

But also, in both the United States and Australia, the *voluntary* take-up rate for insuring dwellings for fire is exceptionally high. In the United States, from 2011 to 2018, only 59%–66% of homes had a mortgage or line of credit secured by a home (averaging 63%) (United States Census Bureau 2021). Meaning 73.5%–87.8% of homeowners in the United States who have a choice, choose to have dwelling insurance coverage for fire. Similarly, the most recent data from the Australian Government is that 53.7% of homeowners have a mortgage (Australian Institute of Health and Welfare (AIHW) 2020). Meaning 76.2%–91.4% of homeowners in Australia who have a choice choose to have dwelling insurance coverage for fire.

One striking feature of these figures is that the data highlights that for homeowners, there is something different about fire risk in particular, in contrast to flood risk. In the United States, only 13%–15% of owner-occupied homes are insured for flood and for 40% of these homes, flood insurance is required, meaning in the United States only 8.2%–9.6% who have a choice, choose to have flood insurance (III 2016, p. 5; 2021c; Strochak et al. 2018). Australia appears to have a somewhat better penetration of flood cover than the United States, but certainly still nothing like the prevalence of fire cover. In 2008, the Institute of Australian Actuaries reported that insurance for the 'overflow of rivers and creeks following long duration rainfall' was 'becoming more common, although is still far from the norm' (Institute of Actuaries of Australia (IAA) 2008, p. 1). By 2011, the Insurance Council of Australia (ICA) predicted that by 2013 flood cover could rise to as high as 30% (Australia Government, The Treasury 2011, p. 22, n. 10).

Why is take-up of fire cover different than flood? It seems to be a combination of two factors. First, in both the United States and Australia there seems to be persistent confusion about whether standard home insurance covers flood (Carter 2012, p. 21; III 2017, p. 2, 6, 9). Second, as the Australia Government describes, the core problem is, 'all home insurance policies include cover for bushfire, earthquake, cyclone and storm, but not flood. … flood cover has traditionally been excluded from home insurance policies, and only over the last decade has flood cover been made available by a limited number of insurers. Where it is available, consumers are often able to opt-out of flood cover and evidence indicates that, when able to opt-out, many policyholders do so' (Australia Government, The Treasury 2011, p. 29). Similarly, in the United States, typically flood cover is excluded from mortgage-required and mortgage-compliant dwelling insurance (FIO 2015,

p. 3). All of this suggests that amongst natural disaster hazards, fire risk to owner-occupied dwellings perhaps is unique in that fire ubiquitously is insured both voluntarily and involuntarily.

This conclusion is bolstered by voluntary take-up rates of fire insurance for renters, which starkly contrasts with take-up rates for fire insurance of the dwelling. A renter, by definition, has no ownership interest in the structure, and so only insures their personal property – the contents of the dwelling – for fire. In Australia, three-quarters of renters do not have personal property insurance (Quantum Market Research 2014, p. 11). In the United States, the take-up rates of renters insurance steadily rose from 29% in 2011 to 57% in 2020 (III 2017, p. 4; 2020, p. 11). For some portion of renters in both Nations, landlords require renter's insurance. At this time, there is no data on what percentage of landlords that is. But whatever the percentage, it means that in both Nations, voluntary take-up of renters insurance for fire loss still materially lags the voluntary take-up rates of homeowners for dwelling loss.

All available evidence suggests that homeowners – and uniquely homeowners – want to insure their dwellings for fire, want to fully insure their dwellings for fire, and think they have fully insured their homes for fire. But they haven't.

The high frequency of inadequate insurance of dwellings for fire in Australia and in the United States

Aspirations aside, while it is hard to know with specificity, it appears that most homeowners are underinsured for a total fire loss, probably profoundly so.

Any discussion of underinsurance begins by discussing how it is even possible. There was a time in the United States when 'Guaranteed Replacement Coverage' (GRC) – what in Australia is called 'Total Replacement' – was ubiquitous. The standard in the United States today is 'Full Replacement Coverage' (FRC) – what in Australia is called 'Sum Certain.' Under GRC, if a covered causal event results in a total loss, then the cost to rebuild is covered, regardless of the cost. Under FRC, by contrast, there is a stated coverage limit, which can be increased through the purchase of 'Extended Replacement Coverage,' but either under FRC or under FRC plus an extension, there is a hard cap. If there is a hard cap, then there is the possibility of underinsurance, meaning the amount of insurance proceeds is inadequate to rebuild the lost home in the event of a total loss.

Work done in Australia illustrates the challenge in answering the simple question: What percentage of homeowners have inadequate insurance to rebuild their homes? In 2005, in the wake of the 2003 Canberra fires, when summarising the Australian research, the Australian Securities & Investment Commission (ASIC) decided to investigate the causes of underinsurance and reported that the percentage of homeowners underinsured by 10% or more had been calculated twice – once as 27.5% and once as 81% (ASIC 2005, pp. 13, 15). ASIC did not opine which number more approximated the truth. Nor to date has any further work been published doing so.

Work in the United States has fared little better. A variety of post-disaster surveys of underinsurance have been done. The consumer advocacy group, United Policyholders, has done a number of post- disaster surveys, finding a range of underinsurance frequencies but generally finding it to be over 50% (United Policyholders survey 2021). One of the early pioneers studying underinsurance, Peter Wells, reported (without transparency as to data or methodology) on his calculations of underinsurance nationwide in the United States over several years (Wells 2007, p. 46). And after the 2008 California wildfires the California Department of Insurance (CDI) performed a Market Conduct Study on underinsurance rates, finding it to be to approximately 80% (CDI 2010, pp. 1027–1030).

All of the analysis from Australia and the United States shares one feature – as a stand-alone data point it is subject to critique. Either the analysis is not transparent and replicable, or it is too focused on a specific region in a specific context, or it is contradicted by other contemporaneous work. Yet thought of collectively, the work tells a story. Figure 9.1 is a chart of every extant, public-facing assertion of underinsurance either in the United States or Australia, whether regional or national, whether post-disaster or not:

This chart simultaneously is frustrating and illuminating. The chart highlights how little currently can confidently be known in any granularity

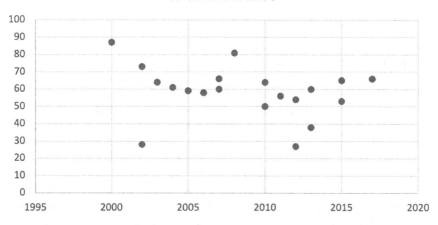

Figure 9.1 United Policyholders

Source: United Policyholders (2021); Fried (2017); Administrative Rulemaking File (2015, pp. 1027–1030); Wells (2007, p. 46); ASIC (2005)

about the pervasiveness of underinsurance. But the chart also makes plain that whatever is going on, it likely is alarming. The possibility of underinsurance seems to be ubiquitous. In other words, most insureds have a coverage limit, and an inadequate one. Because if most homeowners, or even many homeowners, had GRC, then one wouldn't see these high percentages of underinsurance. It simply would not be possible.

One additional data point that is not in this chart is instructive. It is a rate filing by a California insurer after the 2017 Santa Rosa fires (CSAA Insurance Exchange 2018). In the United States, many states require that in order for an insurer to sell a policy at a particular rate, the insurer must file the proposed policy and rate with the State and get state approval (FIO 2015, p. 15). California is one such state (CDI n.d.). The filing by insurer CSAA sought to support a rate increase by detailing the prior claims adjustment experience. And the data provided is consistent with two-thirds of insureds being underinsured, most of them profoundly (by at least 20% or more).

And when one asks, when a homeowner is underinsured, is it by a lot or a little, the news does not get better. The data is sparse. But when the studies go the next step and inquire – when one is underinsured, on average how much is the protection gap? – the answer again is by a lot, likely conservatively on average by at least 20%, more likely by quite a lot more (Klein 2019, pp. 46–50).

So, how does this happen? Most homeowners put little thought into coverage limits. In Australia many homeowners do not seek input. Surveys in Australia suggest 56% of homeowners pick coverage based on their own estimate or the purchase price (which 85% of the time is more than two-years stale), while only 22% rely on the insurer's guidance (Quantum Market Research 2014, p. 17). In the United States, a homeowner simply contacts an agent or broker, goes to a website, or calls a national telephone number, and asks for insurance on their house. The homeowner is asked a few questions, is quoted a policy cover and cost, and a deal is struck.

For the homeowner who wonders whether the cover is adequate, the advice they are given is to ask the insurer. For example, in the United States, the Insurance Information Institute says, '… your insurer will provide a recommended coverage limit for the structure of your home….' (III 2021a; 2021b). The National Association of Insurance Commissioners says, 'Your insurance agent usually will help you decide how much dwelling coverage to buy when you first get homeowners insurance. Your coverage should equal the full replacement cost of your home' (National Association Insurance Commissioners (NAIC) 2010). The website of the Texas Department of Insurance says, 'Your insurance agent can help find out your rebuilding cost.' (Texas Department of Insurance 2021). And the website of the North Carolina Department of Insurance says, 'Below are a few important questions that everyone should ask their agent when purchasing homeowners insurance 1. Do I have enough insurance to rebuild my home if it is destroyed?' (North Carolina Department of Insurance 2022). Australia is no different. In Australia, 'Consumers generally need specialist assistance to estimate rebuilding costs, but it is often impractical to refer to builders, architects or quantity surveyors. Many insurers now provide consumers with access to web-based calculators.' (ASIC 2007, p. 10). Or more simply put, most of the time the estimate of the rebuilding cost of a house, if there is one, comes from the insurer. If the estimate is wrong then the insurance is wrong.

But that does not, in and of itself, explain underinsurance. Because there is no precise, mathematical, objective cost for the rebuilding of a home. Until a home is actually being built, it always is an estimate that will to one degree or another, in one direction or another, be wrong.

One would expect, however, that estimates of coverage to break evenly high and low, meaning the rate of underinsurance should group around the 50% line, and distribute evenly above and below. And it does not. Figure 9.1 illustrates that there are dramatically more instances of underinsurance than of over-insurance.

One way to understand underinsurance rates disproportionately clustered above the 50% line is to think of them as akin to the idea in mathematics called a mathematical fallacy. In broadest terms, a mathematical fallacy is when the conclusion of a proof suggests that there is a flaw in the proof, even if the flaw cannot be identified. Think, for example, of a coin flipped 1000 times. 700 times it comes up heads. 300 times it comes up tails. The experiment is repeated. Now the results are 650 heads, 350 tails. It is repeated again. 800 heads. 200 tails. Something is wrong. Maybe the coin is weighted unevenly. Maybe something else is going on. But it bears investigation. Because something may be amiss.

What may be amiss with dwelling insurance? Why are more homes underinsured than over-insured?

In 2007, ASIC reported, 'Even if a consumer correctly estimates what it would cost to rebuild their home in a one-off total loss, it is almost impossible to know what it will cost to rebuild a home that is destroyed in a mass disaster. The surge in building prices that occurs after a mass disaster can be very unpredictable.'(ASIC 2007, p. 13). This idea – demand surge – also is proffered in the United States (Klein 2019, pp. 69–71). The premise of the proffered explanation –'the surge in building prices that occurs after a mass disaster can be very unpredictable' – bears further study. Catastrophe modelers (creating data streams for vendors who sell costs estimators) contend they *can* predict natural disaster with granularity down to a specific home address (Raizman & Pratt 2021, 1:12;10–1:31:55).

Whether this granularity of modelling is real or not, however, the data suggests that demand surge alone is an inadequate explanation. In the wake of the 2017 Tubbs Fire in California, CoreLogic studied demand surge and found it averaged 15%–30% (Kopperud 2019). In a Market Conduct Study in 2010, the CDI found, however, that approximately 57% of homes that had purchased an extension of their full replacement coverage still were underinsured (CDI 2010, pp. 1027–1030).

If ubiquitous underinsurance is not primarily driven either by demand surge or homeowner choice, then that suggests a new hypothesis: the likely, primary cause of ubiquitous, unintended underinsurance is the estimating tool. Coverage limits are not plucked out of the sky. Rather, insurers estimate rebuild costs using tools in the United States called 'component cost estimators' and in Australia called 'elemental estimating calculators.'

These estimating tools essentially are big data analytics at their finest. Data sets of millions of construction projects and price lists are broken down into individual labour and materials line items, sorted by location and date. These data sets are updated at least quarterly, sometimes more frequently, to capture localised construction cost trends. An insured house is identified as to its elements or components, down to screws and bolts, and then an estimate is built up for the price of building that precise house in that precise location. And, as alluded to above, because of the prevalence of demand surge, other data streams also are involved.

What appears to be happening is that the cost estimating tools informing coverage limits are more often than not simply under-estimating reconstruction costs. Of some frustration is that this is a testable hypothesis, but it has not been tested. There is a moment in time when one knows with precision the cost of building (or rebuilding) a home. That is the moment that construction is actually completed on that home. And at that moment, the home usually is insured. So, at that moment, the estimating tools are deployed. The insurer may not know the actual construction costs of every newly constructed home the insurer insures. But at a minimum, if the home is one that was rebuilt after a total loss, and the rebuild was adjusted under the same insurer that now will insure the new home, the insurer has or had access to the actual cost of building that specific home, which, in turn, means the insurer has a data point allowing it to compare at the same point in time the actual cost of constructing a specific home and the estimated cost to build the same home. A large insurer has access to lots of these data points. Which means an insurer can construct a mature data set from which an insurer can know for its insureds and its estimating tools the frequency of inaccuracy, and the average depth of inaccuracy.

All of this raises two questions: (1) Why would an insurer be open to selling less insurance than a homeowner wishes to buy? (2) Why would an insurer be incurious about how well its estimating tools are working? Each answer is intertwined with the other.

The legal landscape of insurance is complicated, in part because there is not consensus on what precisely insurance is. Is insurance a quasi- public utility to have maximum risk spreading, or is it a variant of a personal security product or is it an ordinary contract? Is the relationship between insurer and insured arms-length, adhesive, or fiduciary? Should insurance markets be free market structures, lightly regulated, or highly regulated? Is insurance a luxury or a necessity? For an insurer, a complex legal and conceptual landscape creates a set of market incentives that may at first glance be counter-intuitive because it creates an incentive for insurers to – in the context of a single customer – sell less of their product.

Whether they are correct or not, insurers perceive their customer as highly price elastic. And while coverage limits pale in comparison with deductibles as a price determinant, coverage limits do change premiums at least on the margin. In other words, an insurer might conclude that in a highly competitive market with what the insurer perceives as highly price elastic customers, there is market share to be gained by a slight drop in price. And that the downside risk is minimal. Because claims in excess of coverage limits – even suppressed coverage limits – are rare. When those claims are adjusted, in many instances the insured will not challenge the adjustment. When the insured challenges the adjustment, many of those challenges will be settled at a discount. And of those that are not settled, the muddled legal landscape will result in a total or partial victory for the insurer much of the time.

But that muddled legal landscape may not work so well for an insurer if the insurer knows (or is found to be wilfully ignorant of) the (in)accuracy of its own rebuild estimating tools. If an insurer knows about the average error rates of its own estimating tools, then it cannot so easily prevail on a position that the homeowner knowingly contracted for less than full insurance. Until the data is collected, it is impossible to know what it will show. But it is reasonable to speculate that if the data was *good* for the insurer then the insurer would tout it. And if the insurer *thought* the data would be good for the insurer, then the insurer would do the research. The fact that no insurer is touting its data is suggestive. But it is far from definitive. Because in both Australia and the United States, while getting the cover correct requires time and expertise, the onus of error falls on the homeowner (ASIC 2007, p. 7; Klein 2019, pp. 82–97).

There may well be solutions to underinsurance. Two possible solutions are to require insurer transparency about the accuracy of its estimating tools, and/or to require – with clarity – insurers to bear the responsibility for understated coverage limits. But solving underinsurance will do nothing to resolve the increasing unavailability of affordable fire insurance. To the contrary, resolving underinsurance could cause prices to rise.

The emerging challenges of affordability and availability of fire insurance of dwellings in Australia and the United States

While presently – in both Australia and the United States – most homeowners either voluntarily or involuntarily do insure their dwellings for fire, that may not be the state of matters for long. Increasingly, both Nations face issues of availability and affordability.

There is inadequate academic literature defining the precise parameters of affordability and availability of fire insurance of dwellings – either in Australia or in the United States – in the face of the increasing frequency and economic impacts of fires. That said, in both Nations the issues are considered emergent. For example, on 8 February 2021, the ICA announced that following 'A range of inquiries and reviews over the past decade focused on issues of insurance affordability and availability in high risk areas or sectors and have identified potential coverage gaps for some groups of consumers and businesses,' the ICA 'is undertaking a review of the insurance sector's options for reforms to improve its contribution to national economic recovery and growth, amid concern from insurers, stakeholders and the community' (ICA 2021). Similarly, on 19 October 2020 the California Insurance Commissioner convened 'an investigatory hearing to initiate a series of regulatory actions that will protect residents from the increasing risk of wildfires. ...to stabilize the insurance market while protecting lives and homes, reducing catastrophic wildfire losses, and increasing transparency for consumers' addressing 'issues including...[i]nsurance availability and affordability' (CDI 2020).

The shape of the problem is not hard to understand. Insurers are profit-seeking businesses. Insurers will only write on homes and in communities that are profitably insurable. Insurability requires risks must be random, well-enough understood to make pricing and underwriting possible, diversifiable, and exist in markets with low levels of moral hazard and adverse selection (Kousky & Light 2019, p. 355). The market behaviour of insurers is consistent with fire cover of dwellings increasingly not meeting this standard.

Put another way, as the frequency, intensity, and economic impacts of fire grow, the affordability and availability of fire insurance shrinks. This is inevitable in a market where insurance is unregulated or lightly regulated. As described earlier, insurers have access to sophisticated data analytics tools that allow them to know with increasing confidence both the likelihood of a wildfire or bushfire coming to an individual home address, and the likely economic impact on that structure if it does so. An insurer will be uncompetitive if it does *not* use this data to isolate high-risk addresses and then either decline to offer cover to those addresses or separately cover those addresses priced in high-risk pools. Because a competitor undoubtedly will, and thereby price cut an insurer that doesn't.

The consequence of this insurer behaviour is pressure on governments to step in with public insurance products as insurance of last resort, or publicly subsidise private insurance products, or step in post-disaster to recompense the losses of the uninsured. A government that creates public insurance products faces the challenge that it may be politically unpalatable to price this insurance as a high-risk pool but may be fiscally reckless not to do so. A government that subsidises private insurance is creating an externality to market forces that both may drive up price and may be politically exposed for the implicit social equity choices embedded in any subsidy program. A government that repetitively steps in post-disaster may create moral hazard behaviours amongst homeowners that reduce take-up rates of insurance – homeowners are less likely to insure large but unlikely losses if they expect that if the loss occurs, then the government will bail them out.

There may be a fourth option. It may be possible to restructure markets to address affordability and availability. This approach would be to remove the vulnerability to price undercutting of an insurer who does not isolate high-risk addresses to an insurer who does. How does one do this? Perhaps by regulation that does not allow ratemaking that accounts for the address of a home. Such a regulation would impose undistorted risk pools, or at least undistorted by property address.

Insurers still could and would rate risk on other vectors. Such as the materials a home is built with. Or whether the home has defensible space around the home. Or the wiring in the walls and its profile for fire risk.

One would expect insurers to still write the risk. Because there are a host of high-value homes in fire-exposed locations, and insurers will not simply walk away from insuring Bondi Beach, NSW, Australia or Malibu, CA, United States.

One would expect the price to be affordable. While not a perfect analogue, this has been the experience of a similar structure in private healthcare insurance in the United States under the Affordable Care Act. And while the data is sparse, there is some reason to believe that in the United States, the total cost of even private *All Perils* insurance would be less than 2% of an average homeowner's annual expenses.

But the data and research are sparse. None of this at present can be known with confidence.

Conclusions

Both the governments of Australia and the United States identify a high prevalence either of uninsureds or underinsureds as concerning (FIO 2015, p. 3; ACCC 2020, p. vii; ASIC 2007, p. 2). The title of this chapter poses a question: Is fire insurable? The answer this chapter gives is unsatisfactory: Perhaps.

This chapter tackles only one issue related to insurance responses to wildfire and bushfire – insurance for rebuilding a home. And this chapter does so largely in cursory fashion. It is important for any researcher to realise that this chapter is only a tree in a larger forest. There is so much more involved in a mass fire event: insurance of out buildings and other structures, personal property, and alternative living expenses; insurance for renters, businesses, farms, and mobile homes; and insurance to mitigate the impacts of displacement of persons and jobs. The policy ripples of fire beyond insurance are mind-boggling. Just to mention a few: relief for the uninsured, health risks, disproportionate impacts on minority communities or across gender lines, mitigation, resiliency, energy policy, climate change, and environmental degradation. Insurance of fire is a book. Fire is a multi-volume set. But ideally this chapter is a primer.

References

Administrative Rulemaking File for Cal. Code Regs, Tit.10, § 2695.183 2015, Ass'n. of Cal. Ins. Cos. v. Jones, 235 Cal. App. 4th 1009 (No. B248622).

Australia Government, The Treasury 2011, *Natural Disaster Insurance Review. Inquiry into flood insurance and related matters*, <https://treasury.gov.au/sites/default/files/2019-03/p2011-ndir-fr-NDIR_final.pdf>.

Australian Competition & Consumer Commission (ACCC) 2020, *Northern Australian Insurance Inquiry*, <https://www.accc.gov.au/system/files/Northern%20Australia%20Insurance%20Inquiry%20-%20Final%20Report%20-%2030%20November%202020.pdf>.

Australian Institute of Health and Welfare (AIHW) 2020, *Home Ownership and Housing Tenure*, <https://www.aihw.gov.au/reports/australias-welfare/home-ownership-and-housing-tenure>.

Australian Securities & Investment Commission (ASIC) 2005, *Getting Home Insurance Right: A report on home Building underinsurance*, Report 54, September 2005, <https://download.A.S.I.C.gov.au/media/1348214/underinsurance_report.pdf>.

Australian Securities & Investment Commission (ASIC) 2007, *Making Home Insurance Better*, Report 89, January 2007, <https://download.A.S.I.C.gov.au/media/1338332/home-insurance-report-Jan-2007.pdf>.

Booth, K & Tranter, B 2018, 'When disaster strikes: Under-insurance in Australian households', *Urban Studies*, vol. 55, no. 4, pp. 3135–3150.

CalFire 2021, *Incidences*, <https://www.fire.ca.gov/incidents/2020/>.

California Department of Insurance (CDI) 2010, *Administrative Rulemaking File for Cal*, Code Regs, Tit.10, § 2695.183 at 1027–1030, Ass'n. of Cal. Ins. Cos. v. Jones, 235 Cal. App. 4th 1009 (2015, (No. B248622).

California Department of Insurance (CDI) n.d., *Rate Filing Review Process*, <http://www.insurance.ca.gov/0250-insurers/0800-rate-filings/rate-filing-review-process.cfm>.

California Department of Insurance (CDI) 2020, *News Release: Advisory – Commissioner Lara holds investigatory hearing on insurance availability and affordability following wildfires*, <http://www.insurance.ca.gov/0400-news/0101-advisory/Advisory-008-2020-investigatory-hearing.cfm>.

Carter, R.A 2012, 'Flood risk, insurance and emergency management in Australia', *The Australian Journal of Emergency Management*, vol. 27, no. 2, pp. 20–25.

Collier, B.L & Ragin, M.A 2019, 'The influence of sellers on contract choice: Evidence from flood insurance', *The Journal of Risk and Insurance*, vol. 87, no. 2, pp. 523–557.

Cronkite, D 2016, 'Force-placed insurance: The lending industry's dirty little secret', *Chicago-Kent Law Review*, vol. 91, no. 2, pp. 687–709.

CSAA Insurance Exchange 2018 *2017 North Bay Fire Claims*, SERFF filing ID WSUN-131288886, <https://filingaccess.serff.com/sfa/search/filingSummary.xhtml?filingId=131288886>.

Federal Insurance Office (FIO) 2015, *Report Providing an Assessment of the Current State of the Market For Natural Catastrophe Insurance in the United States, U.S. Department of the Treasury*, <https://asfpm-library.s3-us-west-2.amazonaws.com/NFIP/DeptTreasury_Current_State__Market_for_Natural_Catastrophe_Insurance_2015.pdf>.

Fried, C 2017, *Recent Disasters are a Wake-Up Call to Check Your Homeowner Insurance*, CNBC, <https://www.cnbc.com/2017/09/05/harvey-is-a-wake-up-call-to-check-your-homeowners-insurance.html>.

Institute of Actuaries of Australia (IAA) 2008, *The Insurance of Flood Risks*, <https://actuaries.asn.au/Library/GIS08_Plenary%204_Paper_Flood%20working%20group_The%20Insurance%20of%20Flood%20Risksl.pdf>.

Insurance Council of Australia (ICA) 2021, *New Release: Insurers to Examine Coverage Gaps to Better Support Economic Growth*, <https://www.insurancecouncil. com.au/assets/media_release/2021/210208%20Insurers%20to%20examine%20 coverage%20gaps%20to%20better%20support%20economic%20growth.pdf>.

Insurance Information Institute (III) 2016, *How Many Homes Are Insured? How Many Are Uninsured?*, blog, 29 January 2016, The Triple-I Blog, <https://www.iii.org/ insuranceindustryblog/how-many-homes-are-insured-how-many-are-uninsured/>.

Insurance Information Institute (III) 2017, *2016 Consumer Insurance Survey, Homeowners Insurance: Understanding, Attitudes and Shopping Practices*, <https:// www.iii.org/sites/default/files/docs/pdf/pulse-wp-020217-final.pdf>.

Insurance Information Institute (III) 2020, *2020 Triple-I Consumer Poll*, <https:// www.iii.org/sites/default/files/docs/pdf/2020_triple-i_consumer_poll_091620.pdf>.

Insurance Information Institute (III) 2021a, *How Much Homeowners Insurance Do I Need?*, <https://www.iii.org/article/how-much-homeowners-insurance-do-you-need>.

Insurance Information Institute (III) 2021b, *Insurance for Your House and Personal Possessions*, <https://www.naic.org/documents/prod_serv_consumer_guide_home. pdf>.

Insurance Information Institute (III) 2021c, *Facts + Statistics: Flood Insurance*, <https://www.iii.org/fact-statistic/facts-statistics-flood-insurance>.

James, P.S 1954, 'Nicholas Barbon—Founder of modern fire insurance', *The Review of Insurance Studies*, vol. 1, pp. 44–47.

Klein, K.S 2019, 'Minding the protection gap: Resolving pervasive, profound, unintended homeowner underinsurance', *Connecticut Insurance Law Journal*, vol. 25, pp. 34–111.

Kopperud, G 2019, *How Demand Surge after Natural Disasters May Impact the Cost and Timing of Recovery*, <https://www.linkedin.com/pulse/how-demand-surge-after-natural-disasters-may-impact-cost-guy-kopperud/?articl eId=6590315770293600256>.

Kousky, C & Light, S.E 2019, 'Insuring nature', *Duke L. J*, vol. 69, pp. 323–376.

Kunreuther, H 2018, 'All-hazards homeowners insurance: Challenges and opportunities, *Risk Management and Insurance Review*, vol. 21, pp. 141–155.

National Association Insurance Commissioners (NAIC) 2010, *A Consumer's Guide to Home Insurance*, <www.naic.org/documents/consumer_guide_home.pdf>.

North Carolina Department of Insurance 2022 *Questions to Ask Your Agent*, <https:// www.ncdoi.gov/consumers/homeowners-insurance/questions-ask-your-agent>.

Quantum Market Research 2014, *Understand Home Insurance Research Report*, <https://understandinsurance.com.au/assets/research/ICA%20Understand%20 Home%20Insurance_Report.pdf>.

Raizman, D & Pratt, S 2021, *FHFA Public Listening Session on Climate and Natural Disaster Risk Management at the Regulated Entities*, YouTube, 19 January 2021, Federal Housing Finance Agency, <https://www.fhfa.gov/Videos/Pages/FHFA-Public-Listening-Session-on-Climate-and-Natural-Disaster-Risk-Management-at-the-Regulated-Entities.aspx>.

Strochak, S, Goodman, L & Zhu, J 2018, 'Too many homeowners lack flood insurance, but many buy it voluntarily', blog, 18 September 2018, Urban Wire: Housing and Housing Finance, <https://www.urban.org/urban-wire/too-many-homeowners-lack-flood-insurance-many-buy-it-voluntarily>.

Texas Department of Insurance 2021, *Homeowners Insurance Guide*, <https://www. tdi.texas.gov/tips/enough-home-insurance.html>.

United Policyholders 2021, *Roadmap to Recovery Surveys*, <https://www.uphelp.org/roadmap-recovery-surveys#a>.

United States Census Bureau 2021, *American Housing Survey Table Creator*, <https://www.census.gov/programs-surveys/ahs/data/interactive/ahstablecreator.html?s_areas=00000&s_year=2019&s_tablename=TABLE1&s_bygroup1=1&s_bygroup2=1&s_filtergroup1=1&s_filtergroup2=1 >.

Wells, P 2007, *Insuring to value: Meeting a critical need*, 2nd ed. Cincinnati: The National Underwriting Company.

10 Fire insurance and the 'sustainable building'

The environmental politics of urban fire governance

Pat O'Malley[1]

In 2017, a fire destroyed the Grenfell Tower apartment building in London, killing 72 people and generating global concerns over building fire safety. In practice, concern had been growing internationally for some time about the fire safety of building cladding, which eventually was determined to be the principal factor in that disaster. Similar fires had been recorded in the recent past such as 'The Torch' building blazes in Dubai, both in 2015 and 2017, the Sharjah fire in the UAE in 2012, and the Lacrosse apartment building in Melbourne. Almost immediately after Grenfell, enquiries were established around the world to investigate the issue. In Australia, for example, the Victorian Cladding Task Force Interim Report (2017) was fairly typical in focusing on the regulation of the problematic cladding materials themselves. Despite the concerns with cladding, the Task Force surprisingly reported without comment a 'widely held view that combustibility standards (of building materials) are too onerous and stifle new product innovation.' At the same time, the report recognised that performance-based fire engineering standards did not lay down the law about materials used in construction, since performance-based engineering:

> permits a fire safety engineer to develop a design, involving the use of *combustible cladding*, that meets the performance requirements of the BCA (Building Code of Australia). In other words, that the use or design of a fire protection system's facade would otherwise achieve the same 'performance standard or outcome' as a non-combustible facade.
> (Victorian Cladding Task Force Interim Report 2017, p. 11, emphasis added)

The report went on to recommend modification of aspects of the performance-based model and its standards. What is not obvious is that the report's acceptance of combustible cladding – provided that it met performance standards – reflected the preferences of one side in a politics of fire prevention that stretches back many years. In recent times this politics has centred the environmental concept of sustainability. While this may appear as progressive in an environmentally conscious world, it will be shown that

DOI: 10.4324/9781003157571-13

almost everything about sustainability is contested, including how it is measured and the tools for assessing it – albeit under a veneer of general acceptance that 'sustainability' is a focal concern. This contest – reflected in 'fire politics' around the English-speaking world at least – pits developers and some government agencies against fire insurers and fire departments. Crudely put, developers and their allies seek to minimise the costs of fire prevention. In pursuit of this goal, they support: performance-based (rather than prescriptive) standards where these allow for cost reductions; a focus on the building's fire preventive sustainability to the point of transfer to the owner ('cradle to gate'); and emphasising 'active' fire prevention (such as sprinklers) as means for reducing costs. On the other side, insurers and their allies focus on the total life course of the building and seek to maximise long-term fire prevention ('cradle to grave'). Among other things this means putting greater emphasis on 'passive' fire protection (materials and construction), stressing prescriptive standards, and highlighting the environmental costs of fire damage and reconstruction. Both parties, in turn, seek to define, measure, and operationalise 'sustainability' in ways that mesh with these contrasting interests and strategies.

The report by the Government of Victoria thus appeared as a signal victory to the developers as it required little change to their existing practices and confirmed the status of performance over-prescriptive standards. Yet the report also noted that insurers were using the insurance premium to enforce compliance to their preferred standards – by raising premiums or refusing to insure where these were not adhered to. This has been a long-term tactic of fire insurers around the world, indicating that the fire politics continues, whatever the report may suggest, and whatever apparent agreement there is about the importance of environmental sustainability.

While this chapter will focus on the struggle as it has been played out in Australia, it is by no means an isolated case study. Rather, as mentioned, it is one site of a struggle carried on over the long term and internationally (e.g. Tebeau 2003, O'Malley & Hutchinson 2007). And, as will be seen throughout this chapter, the politics has been fought out in and through international technical literature – as research is carried out by both sides in the struggle, and as independent research is appropriated by whichever side regards findings advantageous to its interests. The fire politics in Australia has echoed, drawn upon and deployed arguments, tools, techniques, and so on that originate overseas – particularly in North America. As well, as may be expected, parties to the struggle in Australia are often components of international conglomerates, particularly in the building and insurance industries[2].

Fire insurance, fire politics, and sustainability

For the better part of two centuries, the fire insurance industry internationally has been a major protagonist in politics focused on preventive policies and practices aimed at protecting the built environment and its inhabitants from fire. In both North America and the United Kingdom, throughout the 19th century, fire insurers pioneered the use of systematic inspection and data gathering aimed at identifying fire risks, especially in industrial settings. (Tebeau 2003; O'Malley & Hutchinson 2007). Building up a considerable data base, and deploying actuarial practices standard in the wider insurance industry, fire insurance greatly influenced the early formation of a risk-based regime that has substantially shaped the urban fire environment through, *inter alia*, building regulation, infrastructure planning, industrial standards, and public safety measures. Indeed even into the twentieth century, as in Australia for example (O'Malley & Roberts 2014), the insurance industry effectively became the principal regulatory agency for urban fire prevention where state intervention lagged. This was something core to fire insurance industry practice, as from the mid-1800s, across the English-speaking world at least, insurers had set fire safety standards and enforced these through the threat of refusal to provide insurance for non-compliance and through the use of premium pricing to drive improvement in fire protection.

Despite this, there has never been a clear and consensual path towards urban fire security even while all parties would subscribe to the ideal of a fire-safe environment. Rather, the field has been one of more or less continuous contestation. For obvious reasons, fire insurers overwhelmingly have been on the side of maximising fire security. While it could be argued that a fire-secure environment would destroy the market for fire insurance, insurers have both an interest in keeping premiums affordable, and in protecting insured properties against fire risks from adjacent buildings. Fire services – often founded by and still often partially funded through fire insurance – have been a consistent ally, focusing on the lives of their members and the public. But property developers, local governments, and other industrial interests have frequently opposed or passively resisted increased fire security, largely because of the often considerable expense involved. This 'fire politics' has been a central feature of international urban fire security more or less continuously since the 1850s (O'Malley & Hutchinson 2007).

Given the general prominence of sustainability rhetoric in the public sphere, it is not surprising that recently urban fire politics has been considerably shaped by a concern with environmental sustainability. Both in terms of the resources consumed to create fire safety, and the resources destroyed in fires, sustainability in this field has focused on the environmental implications of buildings: summed up in the term 'the sustainable building.' This intersection between sustainability and fire safety is much more than merely a side issue in contemporary urban fire governance, and despite superficial appearances sustainability is far from an agreed-upon goal. Instead, sustainability has become the site of a core struggle, in which the fire insurance industry again has been pivotal. On the surface it would appear as though the key contenders in this politics have harmoniously adopted the practice of sustainability, and are collectively advancing the building regulatory reform agenda towards a shared objective of greater environmental sustainability. This, however, conceals a much more complex reality. The discourse of sustainability – and indeed the very methodologies whereby it is translated into practice – have been co-opted and employed by the various parties to advance their traditional political and economic interests.

Defining sustainability

The political contest begins at the definitional level for there is no generally accepted definition of sustainability in fire politics. The apparent absence of conflict here is an effect of the fact that the key players operate with different definitions of sustainability, such that all can profess to be advancing the spirit of sustainability. In practice, the various major players have promoted conceptions of sustainability that embody and further their respective long-term interests concerning fire safety. The same holds for those terms associated with the concept of sustainability, chief among them 'green building' (as both an object and activity), 'environmentally friendly building' and 'eco-friendly building' (Tidwell & Murphy 2010, p. 2; Merrill Lynch 2005, p. 1). An analysis of fire politics must therefore unpack the various definitions of sustainability (and its associated discourses, concepts, and methodologies) that express the various players' concerns. As will rapidly appear, both fire insurers and their opponents render sustainability an economic issue, and in so doing translate the politics of sustainability into a continuation of a struggle that has characterised fire insurance and fire regulation for over a hundred years. Perhaps unexpectedly, a focus on linking fire prevention and environmental sustainability began not with fire insurers, but with property developers.

Property developers and the economisation of sustainability

At the end of the 20th century, property developers recognised the potential of environmental sustainability to produce profit. Until then, the assumption had been that a green building was simply a luxury for which clients paid a premium and which thus drove down overall demand. But as environmental concerns became widespread, developers re-examined the concept of environmental sustainability, and saw less of a conflict between market and cost considerations and going green. Thus began the 'economisation' of the concept of sustainability, that is the deployment of discursive practices that render any phenomenon part of 'the economy' Caliskan and Callon (2009, 2010)[3]. In Australia, leading the way in economising sustainability in fire prevention, was the Investa Property Group, which started the process in 1999 on the assumption that sustainability was good for profit as a marketing tool, as observed by many others since (e.g. Merrill Lynch 2005, p. 8). So often are marketing advantages now given as an incentive for 'going green' that they emerge as a prime mover in the sustainable building movement: going green thus is revealed not as an end in itself but as a major contributor to economic profitability (Carter 2011, p. 54). In a closely linked way, more generally, corporate social responsibility has blossomed in recent times precisely because of this increased recognition that ethical business 'pays.' It is in this context that we can make sense of 'sustainable' fire safety practices emanating from the developers, who – perhaps not wanting to appear economic opportunists – publicly proclaim many other reasons for their alignment with green building. These include pressure from government regulations, the impact of energy ratings, capital expenditures, reduced running costs, and 'ethical reasons' (Carter 2011, p. 62). Each of these however converges with the financial interests of the developers.

While this convergence is more or less self-evident with respect to marketing and reduced capital expenditure, reductions in running costs and improved energy ratings both provide marketing resources while the latter may attract subsidies and reduced costs.

And while ethical concerns may be genuine, these are also major marketing assets. More problematic is the issue of regulation since, as will be seen later, government regulation has only followed developments emanating from industry[4]. The state, following a more or less explicit neo-liberal agenda favouring self-regulation, has not been proactive in advancing the sustainability agenda in the Building Code of Australia (BCA)[5].

How, then, do the developers' fire-safety objectives fit into the picture of sustainability? Here the BCA has been important, since it has promoted performance-based, rather than prescriptive, regulation. Where self-regulation cannot be given free rein, for example where public safety has to be guaranteed, regulation has tended towards regulation of ends (performance regulation) rather than means (prescriptive regulation). In the case of fire protection, this means moving away from *prescribing* how safety is to be achieved, for example by the specification of materials and design of infrastructure, towards giving leeway to innovate how certain specified ends are to be achieved. In this regulatory environment, developers now interpret minimal allowable capital expenditures on fire safety as productive of minimal *embodied carbon* (that is, the amount of carbon emissions required in the production of a good or service). From this perspective stripping buildings of certain *passive* fire protection systems – such as thicker and more numerous walls for greater fire resistance, and increased compartmentalisation – is heralded as sustainable. This preferred fire-safety configuration requires for its construction far fewer carbon emissions than would a more comprehensive passive fire protection strategy[6]. An equivalent level of fire safety is, of course, proclaimed, on the basis of performance-based fire safety engineering methods. Through quantification, precision modelling, and computer simulation, a degree of 'over engineering' (Carter 2011, p. 71) in fire safety is identified, and thereby rendered redundant. Today, through the development of performance-based fire safety engineering, the variables in fire safety assemblages can be so precisely controlled as to allow for substantial reductions in passive fire protection investment. Developers and their engineers understandably consider this a triumph of rationalisation, for economic resources are now argued to be more efficiently distributed. Minimalist fire safety – in accordance with that specified in the BCA – is therefore rational-ised (in both senses of that term) as increasing sustainability.

However, minimising initial construction costs, now rendered as reducing carbon, is not the sole concern of the developers' conception of sustainabil-ity as most of a building's carbon emissions are produced not at the time of construction, but over the course of its operation. Potential owners, seeking to reduce their own costs, prize energy-efficient buildings, and this, in turn, drives demand for sustainable building designs. Not only are buyers and renters attracted to long term energy savings but building owners recog-nise that therefore they can charge higher rent from tenants (Carter 2011, p. v) Through the pricing of carbon, it has become clear to developers that 'a building's operating costs are where the greatest cost savings are to be made' (Gritzo 2009, p. 4).

While it is thus clear that there are economic incentives to developers to 'go green' in doing so they have often had to run against traditional fire protection principles – particularly those concerning passive fire protection. Under previous prescriptive building regulation, these performance-based 'sustainable' designs would not have satisfied the BCA requirements. The sustainable building industry has thus rested on the technological promise of performance-based fire safety engineering. The issues created may be illustrated by the efforts of developers to tackle energy costs by substituting natural for artificial light. The developer can, for example, opt for skylights, larger windows, open floor plans, or atria. All of these present immediate fire safety issues: skylights weaken the structural integrity of ceilings and pose a danger to the firefighters; larger windows may reduce the fire resistance of the glass; open floor plans promote a faster fire spread; and atria pose prob-lems for the vertical containment of fire spread and may delay the activation of smoke detectors. On their own, these problems could decisively condemn sustainable designs. Yet solutions to all of these problems have been devised that depend upon performance-based fire safety engineering to meet the BCA's requirements. Basically, in each example, a finely tuned, computer modelled and simulated interconnected network of fire protection technol-ogies ensures that the various contingencies are taken into account. This involves an assemblage of active fire prevention, including smoke detectors, automatic fire sprinklers, sophisticated smoke evacuation systems, fire cur-tains, fire doors, manual fire suppression devices, etc. These together func-tion in interlocking ways to provide what developers insist is a fail-safe fire safety system. Another example, of course, is the use of such technologies to negate the risks posed by flammable building claddings.

Fire insurers and the economisation of sustainability

Nonetheless, these 'solutions' reliant on active fire prevention have not sat-isfied fire insurers or their allies the fire services, and passive fire protec-tion manufacturers. For example, sprinklers can fail, and when they do, the structural integrity of the building may be put in sufficient doubt to prevent fire brigades from fulfilling their statutory duties. The building's very durability may thus be compromised, with ramifications on life, property, the environment, and business continuation. From this other side of the fire politics, 'sustainability is just an excuse used by the building designers and owners for why fire protection measures are excluded' (Carter 2011, p. 52). A building that is so heavily dependent on sprinklers and other active fire prevention measures, and that has been stripped of those passive fire protection features that would mitigate the damage in the event of sprinkler failure, is a building bearing excessive fire risks. In short, '(e)nergy efficiency measures, a critical component of green construction, may increase the risk of fire substantially' (Tidwell & Murphy 2010, p. 7).

This begins to open out the conflicts concealed behind the apparently clear-cut issue of sustainability. For the fire insurers and their allies there is an alternative conception of sustainability, one that deals in durability, life chances, and resilience. In essence, it is a more generalised definition of sustainability. Thus leading global fire insurer FM Global speaks of 'sustainability in general,' by which is meant 'the protection not only of properties and their assets, but also of people, their livelihoods, the environment, the local community, the economy' (McGrath 2008, p. 3). It is this opposition to identifying sustainability with fire protection minimalism, that distinguishes the insurers' more expansive (but still economised) definition of sustainability.

The insurers' approach to sustainability is just as much predicated on securing economic gain as is that of the developers: insurance after all is the capitalisation of risk (Ewald 1992). *But where developers have an economic focus on reducing the capital expenditures in construction and the operating costs of their buildings, insurers have an economic focus on mitigating the risks that might befall the buildings they guarantee.* To insurers the durability of the buildings in question is therefore of prime importance and like developers the insurers have deployed a green vocabulary to give their traditional economic concerns with property protection an aura of universality. For insurers, only when both passive and active fire prevention measures are optimised, rather than minimised, will a building have reached the apex of sustainability. Clearly, the discourse of sustainability functions here as a discursive means in the pursuit of *very* long established interests.

This is highlighted by the contrast with the concerns of the insurers' key allies the fire services, for whom durability and resilience are similarly central components of their approach to sustainability. While insurers have a shared economic interest with the passive fire-protection industry, fire services are motivated not by economic gain, but by their statutory responsibility to protect life and property, as well as by their concern with the safety of firefighters themselves. The contrast is instructive, but all of this, and the insurer's vision, are brought into sharper focus once we move to the methodological dimension of sustainability.

Calculating sustainability

As with definitions of sustainability there is a *seeming* consensus on the appropriate methodology for ascertaining environmental costs (that is, calculating the level of sustainability). Called life-cycle assessment (LCA), this measures the environmental impact of a product or service throughout the duration of its operation, and is usually contrasted with cost-effectiveness analysis. (Robbins et al. 2009, p. 78). The insurers, fire brigades, developers, and fire protection manufacturers all endorse and employ LCA. However, the politics of sustainability reveal themselves again when focus is directed away from what appears as a common methodological practice, towards the assumptions and inputs that make it up.

At first glance, LCA could appear as anathema to developers' interests since they are solely concerned with the construction of buildings, and not with environmental impacts over their lifetime. However, it has been seen that developers work with an economised definition of sustainability that promotes buildings with reduced operating costs. This renders LCA an attractive methodology for developers, and explains their enthusiastic uptake of a seemingly antithetical means of operationalising sustainability. However, this should not conceal the fact that minimising capital expenditures at the construction stage remain the developers' chief concern, provided all safety regulations are met. Because of this, a specific aspect of LCA methodology lends itself particularly well to the developers, namely: 'cradle-to-gate assessment.'

This is where the environmental impacts are calculated beginning from the point of resource extraction (*cradle*) until when the product arrives at the factory gate (*gate*). Basically, it calculates the environmental impacts of the resource extraction and production phase. Calculation of sustainability is thus based on sustainability costs to the point of delivery. It thus does not incorporate the environmental impacts of the building's *use phase*, nor of its *disposal phase: for example if it burns down and/or is demolished. Cradle-to-gate* assessment is contrasted with *cradle-to-grave assessment*, which is the calculation of environmental impacts throughout the product's life cycle, from resource extraction (*cradle*) to product disposal (*grave*). Carbon emissions in the construction phase are priced, and their minimisation effectively becomes a question of cost minimisation. Cradle-to-gate assessment, in the hands of the developers, thereby is transformed into a methodological justification of their minimal investment in fire safety. Similarly, environmental impacts in the use phase are priced, and so minimisation also emerges as a question of cost minimisation. Cradle-to-grave ecological assessment is thus transformed into a methodological justification of their minimisation of building operating costs. Having economised the concept of sustainability, and having defined sustainability in such a way that it involves reducing initial capital costs and minimising building operating costs, developers have thus merged traditional cost-effectiveness analysis with LCA. Developers, to put it simply, have appropriated an entirely novel methodology for thinking about fire safety precisely because it leaves everything unchanged.

Needless to say, the insurers, fire brigades, and passive fire protection manufacturers all object to what they regard as the developers' exploitation of LCA to pursue traditional cost-minimising fire safety priorities. Essentially, the insurers' argument runs as follows. Minimising both initial capital expenditures on fire safety and building operating costs, as the developers are claimed to be doing, may well reduce carbon emissions. But simultaneously it presents the potential to increase the contribution of fire risk factors to sustainable design. This is because in the event of a fire in a building designed in accordance with developers' preferences for minimal fire safety, the environmental impacts *of the fire itself and of the ensuing rebuilding process end up outweighing the carbon gains presumed by the developers* (Gritzo 2009, p. 13).

Of course, this risk of fire is contested by the developers, especially by pointing to the performance engineering advances in active fire prevention. The point here is not to determine which side is correct, even were that possible, but to note that common adherence to LCA does not imply a shared methodology of sustainability but a contested site of a traditional economised fire politics.

The tools of sustainability

Within the broad conflicting methodologies outlined under LCA, are arrayed specific tools, largely tools of assessment and realisation, that give more precise shape to the respective positions. In 2002 'green' property developers formed the Green Building Council of Australia (GBCA), whose mission 'is to develop a sustainable property industry for Australia and drive the adoption of green building practices through market-based solutions' (Green Building Council of Australia 2002). The GBCA stands as the leading advocate of the green building industry in Australia, and has close ties with the Property Council of Australia (PCA), the peak body for developers. The PCA itself has become involved in the practice of sustainability, and both it and the GBCA endorse the methodology of LCA. Part of the GBCA's advocacy was the development, starting in 2003, of the first comprehensive voluntary sustainable-building rating tool, namely the Green Star rating tool. Subsequent incarnations have appeared in the years since, (the latest version in August 2016) applying to different categories of performance and building types. Both 'resilience' and 'sustainability' appear in the tool but resilience is 'resilience to climate change,' while sustainability focuses on the carbon neutrality in the building and day-to-day operation of the building. (Green Building Council of Australia 2016) In this way, the Green Star rating tool addresses at great length issues of sustainable building with respect to communities, design as built, interiors, and performance. However, the contentious issue of fire safety has been sidelined.

The *insurers* (and their associated engineers such as FM Global) advocate a rather different view of sustainable development. In line with their rendering of sustainability as building resilience and durability, the focus of their sustainability agenda has been on the refinement of the premium – their traditional means of incentivising resilient and durable buildings that factor-in acceptable levels of passive as well as active fire protection (O'Malley & Roberts 2014). The Insurance Council of Australia (ICA) has developed relevant risk databases to aid in this effort, and in particular has been working on a publicly accessible Building Resilience Rating Tool (Edge Environment 2013) to assist in the calculation of resilience to hazards such as fire, and in assessing premiums. Its focus is thus not centred on broader questions of the environment and sustainability as such, but with respect to risks to capital.

Although the common-sense view might see a certain communion of interests between the green property developers and the insurers, it is instructive to point out that there is no substantial mention of fire safety by the green developers. Moreover, that the green developers and the insurers are not on the same page is further evidenced by the important fact that as noted the Green Star rating tool – itself an initiative of the green building industry – has no fire safety criteria. In other words, Australia's leading sustainable building rating tool can deem a building sustainable even as it fails to cater for anything beyond the strict minimum fire safety standards set out in the BCA. With respect to building rating tools, in other words, the insurers and the green building industry are operating with quite divergent understandings of resilience and sustainability. Each has a rating tool to match. This closely maps onto the pattern emerging when consideration is given to the fire protection manufacturing industry.

The *fire protection manufacturers* (the active *and* passive fire protection technology manufacturers) have integrated the discourse of sustainability into their product marketing. Active fire protection manufacturers (allied largely with developers) incorporate notions of sustainability in both their automatic and manual fire-suppressant technology; from sprinkler systems that economise (in both senses) water usage to fire extinguishers with more economical (again, in both senses) release rates and chemicals, sustainable active fire protection translates into cost reductions. As with the developers, this carefully eschews issues of safety and the environmental costs of damage and destruction due to fire. Passive fire protection manufacturers, (allied largely with insurers) on the other hand, incorporate notions of sustainability that now include 'resiliency, life-cycle analysis and occupant health' (San Diego 2013, p. 29). The claim is that passive fire protection 'helps *reduce the loss of* components and materials, with the environmental benefit of reduced resource needs including raw materials, manufacturing energy and resources for construction, reconstruction and renovations required due to fire damage' (San Diego 2013, p. 29 emphasis added).

While its allies in insurance centre resource (and thus economic) issues, the fire services' response is far less concerned with this economisation of fire protection. They understand that the discourse of sustainability, in the hands of developers, reinforces the hegemony of the fire safety standards prescribed in the BCA. Thus in the example of protecting atria and other open spaces, developers promoted introducing sprinklers as a key component of the building's fire safety system. But the sprinklers they settle on are not just any sprinklers: they are sprinklers made more consistent with sustainability because they are sourced with grey water and driven by pumps that optimise water usage through vaporising sprays. The fire services object that these may accumulate chemical deposits sufficient to render them ineffective, and question whether the sprinklers will be powerful enough to douse the incipient fire fast enough before it spreads in a building that has abandoned compartmentation and other trusted passive fire protection measures. In sum, the fire services are concerned that this form of sustainability may impair their ability to carry out their statutory obligations to protect life, property, and the environment – and to carry them out safely.

Conclusions

While it may appear that bipartisan support for the significance of sustainability will lead to a generally accepted, environmentally sustainable regime of fire prevention, this seems to be something a long way off. Expert comments such as '[c]onstructing a building to be more sustainable can sometimes mean making the building less fire safe while making a building more fire safe sometimes requires the building to be less sustainable' (Carter 2011, p. 7), show the grounds for ongoing conflict. But as this chapter has argued, even such cautious comments seem to accept the concept of sustainability itself. They take for granted the existence of a uniform conception of sustainability: precisely that which has been shown to be misleading. A critical examination of the concept and its employment reveals a much more complex and insoluble problem than these comments otherwise suggest.

The persistence of conflicting conceptions of sustainability suggests there is unlikely to be any technological resolution to the struggle because the divide is not just about technological possibilities. It is a political and economic conflict over contradictory fire safety objectives that have long fractured fire prevention. Even so, it may be supposed that the Grenfell disaster has changed things, not only because governments often have had to contribute to the cost of replacing flammable cladding in thousands of buildings. The human cost of the Grenfell tragedy might be seen to trump the economisation of fire safety. And to this development seemingly hostile to developers may be added the influence granted to insurers by their willingness to increase premiums or refuse insurance for buildings that do not satisfy their sustainability standards. Historically this has been a powerful force for change. However, despite their use for well over a century and a half, the coercive potential of premiums has never ended the struggle with developers, owners, and governments. There are limits to the pricing of premiums and competition among insurers has weakened their clout before now. And it was seen earlier that strong support remains for performance-based standards of fire engineering that would permit flammable cladding provided that other engineering measures minimise the risks they pose. The regulatory balance may shift away from the developers for a while, but the long history of fire politics suggests things are unlikely to be resolved by single disasters, of which there have been many. Equally unlikely, given the conflicting discourses of sustainability, is that environmental concerns will prove decisive.

Notes

1 This research was funded by a grant from the Australian Research Council (ARC) Discovery Projects (DP)DP1093187 'Risk, Urban Fire Protection and Security Networks.' I would like to acknowledge the contributions of my two research assistants on this project, Alex Roberts and in particular Alex Lombard who contributed substantially to this paper.
2 This chapter is drawn from two studies of the long-term development of risk-based fire security in Australia and North America indicated in Note 1. This project involved systematic review of the secondary literature on the development of fire prevention, government regulation and fire insurance, and content analysis of news media accounts of urban fire events history, side by side with this work has been a detailed analysis of the records of fire prevention associations, fire services, government agencies, and insurance companies, especially in Canada and Australia, As well, systematic examination has been made of the technical literature that has grown up around what is here referred to as 'fire politics,' and which – while usually objective and scientific in itself – has been deployed by both sides in that politics.
3 For a close analysis of how this has occurred in fire protection, see O'Malley and Roberts (2014).
4 While there has been some local council activism over sustainable building regulations, this has largely been kept in check. Local councils, by virtue of closer proximity to their constituents' building activities, have sought to regulate sustainable building practices at the micro-level. However, consideration of the multitude of widely divergent (and often internally inconsistent) approaches adopted at this level of government is not possible in this chapter.
5 For a full discussion of this see O'Malley and Lombard (2013).
6 Underlying the confidence in the efficacy of these minimalist fire-safety configurations is, of course, performance-based fire safety engineering. It was Vaughan Beck who first pointed out, and advocated, reductions in (claimed to be redundant) passive fire protection in the wake of performance-based fire safety engineering analyses. He specifically noted that performance-based fire safety engineering allowed decreases in the fire resistance of certain passive fire protection features, and that consequently this made possible considerable economic gains, gains that could be invested more rationally elsewhere (Beck 1983, 1987, Beck & Poon 1988).

References

Beck, V 1983, 'Outline of a stochastic decision-making model for building fire safety and protection', *Fire Safety Journal*, vol. 6, pp. 105–120.

Beck, V 1987, 'A cost-effective, decision-making model for building fire safety and protection', *Fire Safety Journal*, vol. 12, pp. 121–138.

Beck, V & Poon, S 1988, 'Results from a cost-effective, decision-making model for building fire safety and protection', *Fire Safety Journal*, vol. 13, pp. 197–210.

Caliskan, K & Callon, M 2009, 'Economization part 1. Shifting attention from the economy towards processes of economization', *Economy and Society*, vol. 38, pp. 369–398.

Caliskan, K & Callon, M 2010, 'Economization part 2. A research programme for the study of markets', *Economy and Society*, vol. 39, pp. 1–32.

Carter, M 2011, *'Promoting the design of buildings that are fire safe and sustainable'*, PhD thesis, A Review for Fire Protection Association Australia, Worcester: Worcester Polytechnic Institute.

Edge Environment 2013, *Insurance Council of Australia flagship project: Building resilience rating tool*, <www.edgenvironment.com>.

Ewald, F 1992, 'Insurance and risk', In Burchell, G, Gordon, C and Miller, P (eds), *The Foucault Effect. Studies in Governmentality, 1991*, London: Harvester/ Wheatsheaf, pp. 197–210.

Green Building Council of Australia 2002, *The Green Building Council of Australia*, <www.gbca.org.au>.

Green Building Council of Australia 2016, 'Rating System', www.new.gbca.org.au.

Gritzo, L 2009, 'The influence of risk factors on sustainable development', *Research Technical Report*, Norwood: FM Global Research Division.

McGrath, B 2008, 'Awash in sustainability', *FM Global Reason Magazine*, vol. 2, p. 3.

Merrill Lynch 2005, 'Green property: does it pay?' Report for the United Nations sustainability project, 6 December 2005, Sydney.

O'Malley, P & Hutchinson, S 2007, 'A genealogy of fire prevention' In G. Brannigan and G. Pavlich (eds.), *Governance and Regulation in Social Life*, London: Cavendish/Glasshouse, pp.145–164.

O'Malley, P & Lombard, A 2013, 'The neo-liberalisation of fire protection in Australian building regulation: The case of performance-based fire safety engineering', *Working Paper 5, Fire Protection and Urban Security Project*, Sydney, University of Sydney.

O'Malley, P & Roberts, A 2014, 'Governmental conditions for the economization of uncertainty. Fire insurance, regulation and insurance actuarialism', *Journal of Cultural Economy*, vol. 7, pp. 253–272.

Robbins A, Wade C, Bengtsson J 2009, 'Sprinklers and sustainability', *Build*, pp. 78–79.

San Diego, D 2013, 'Sustainable design & advanced fire rated glazing in schools', *Private University Products and News Magazine*, May 2013.

Tebeau, M 2003, *Eating Smoke. Fire in Urban America 1800–1950*, Baltimore: John Hopkins University Press.

Tidwell, J & Murphy, J 2010, *Bridging the Gap: Fire Safety and Green Buildings*, Cheyenne: National Association of State Fire Marshals.

Victorian Cladding Task Force Interim Report 2017, <www.planning.vic.gov.au/- data/assets/pdf-file/0016/09412/Victorian-Cladding-Taskforce-Interim-Report- November-2017.pdf>.

Section IV

Air

Always there, it allows itself to be forgotten...
Luce Irigaray

Air, in various manifestations, is an element of ancient China, Egypt, and Greece, and in Buddhism and Hinduism; of atmosphere, breath, life force, weather, and wind. It pervades and intersects bodies, things, and places – sustaining, lifting, warming, cooling, and sculpting. Or, eroding, blasting, freezing, collapsing, and flattening.

The agency of air is taken for granted. In its collusions and hybridizations, as matter it remains invisible and as force its power attracts little more than a casual nod. It is the things and entities that are sustained, warmed, eroded, and collapsed that draw attention; it is a prejudicial primacy of things that attain the status of *matter* of concern (Adey 2015; Jackson & Fannin 2011). Thought and action teeter on the earth's round crust, tethered to and through things, yet seemingly immersed – assembled – in nothing.

Irigaray (1999) argues that there is little benign about this 'forgetting of air.' An airy world is one of flux and change as there is *never* the same air and thus *never* the same place. Supressing this airiness represents a drive for mastery and control, an attempt to fix things and beings. Forgetting air produces a sense of permanence and order that enables assertions of power that foreclose alternate thoughts and actions. Remembering air – proceeding towards re-remembering air – loosens thought and action as this sense of fixity becomes disorientated and doubtful. No place is ever the same place as the air is never the same air.

There appears to be a partial remembering of air in insurance. A preoccupation with the potential loss of things and healthy bodies is recalibrating through the emergence of weather derivatives (Müller et al. 2017). Broadly falling under the banner of climate insurance, derivatives use measured weather events – including wind speed and extreme heat – as triggers for payouts, rather than paying out for actual losses. The uncertain and risky ruses and wiles of this illusive element are, momentarily, rendered tangible and calculable. Things no longer need to be fixed to particular locations – as if there was always the same air, or lack thereof – for insurance to act.

DOI: 10.4324/9781003157571-14

There appears also be a remembering of the flux of air in how households envisage insurance. Moment to moment, insurance shifts in and out of focus and changes shape and form as people describe how insurance may or may not work for them in the event of a natural disaster (Booth & Harwood 2016). Sometimes financial safety network, sometimes moral signifier, the flux and flow of these insurantial moments signpost airy, elemental assemblages of insurance and insurability.

Instead of teetering on a crust surrounded by things, Ingold (2008) envisages a 'weather world' in which all is constituted through and within the binding and unbinding of earth and atmosphere. Thoughts and actions are immersed-emergent in the 'incessant movements of wind and weather' (Ingold 2008: 1803–1804). For example, in megacities, air spells out patterns of inequity: 'air tells us about difference... who belongs and who does not, who is deserving and who is not' (Adey 2013, p. 291), and air is 'socially and bodily significant for shaping our everyday life' (Hauge 2013, p. 171).

Thinking through insurance as part of this atmospheric weather world is the purpose of this section. Airy landscapes, where atmosphere meets and entangles with earth bound objects and entities, demonstrates dimensions of tangibility and intangibility. In the first chapter, Zac Taylor describes how hurricane risk in Miami, Florida is 'exported' to Singapore as the location of risk brokerage. These insurer practices make evident both the globalized force of insurance markets and the specificities and conjunctures of different types of place. The conditioned air in Singapore brokerages and the seasonally devastating winds of the southern United States are intangibly co-constituted. Even in the same place – Cairns in cyclone-prone northern Australia – the air in the form of disaster is varied in its constitution and tangibility. In the next chapter, Nick Osbaldiston describes how some local residents are relatively unphased by cyclonic winds and are uninsured, while for others this fast-paced unpredictable air is a cause of considerable concern particularly in light of escalating insurance premiums.

In the final chapter in this section, Christine Eriksen and Jonathan Turnbull describe how the intangibility of air – specifically air-borne radiation from the Chernobyl nuclear disaster site – contributes to a lack of insurability. The universalizing logic of insurance tends to rely on the placement of discrete objects and entities (even if these are mobile) that are defined and mastered through actuarialism. The limits of this become apparent as radiation is released and travels, unseen, upon the wind. In relation to some air and thus some places, the seeming mastery and permanence of insurance as part of everyday life and in global finances is doubtful.

References

Adey, P 2013, 'Air/atmospheres of the megacity', *Theory, Culture and Society*, vol. 30, pp. 291–308.

Adey, P 2015, 'Air's affinities: Geopolitics, chemical affect and the force of the elemental', *Dialogues in Human Geography*, vol. 5, pp. 54–75.

Booth, K & Harwood, A 2016, 'Insurance as catastrophe: A geography of house and contents insurance in a bushfire prone area, *Geoforum*, vol. 69, pp. 44–52.

Hauge, B 2013, 'The air from the outside: Getting to know the world through air practices', *Journal of Material Culture*, vol. 18, pp. 171–187.

Ingold, T 2008, 'Bindings against boundaries: Entanglements of life in an open world', *Environment and Planning A*, vol. 40, pp. 1796–1810.

Irigaray, I 1999, *The forgetting of air in Martin Heidegger*, London, UK: Athlon Press.

Jackson, M & Fannin, M 2011, 'Guest editorial: Letting geography fall where it may – aerographies address the elemental', *Environment and Planning D: Society and Space*, vol. 29, pp. 435–444.

Müller, B, Johnson, L & Kreuer, D 2017, 'Maladaptive outcomes of climate insurance in agriculture', *Global Environmental Change*, vol. 46, pp. 23–33.

11 The relational urban geographies of re/insurance

Florida hurricane wind risk and the making of Singapore's catastrophe finance hub

Zac J. Taylor

Introduction

In the keynote speech at the 2019 Singapore International Reinsurance Conference, a senior Singapore government minister outlined the city-state's plans to expand as an offshore property catastrophe re/insurance[1] centre (Monetary Authority of Singapore 2019). 'The global economy is undergoing a tumultuous period of change, and facing strong headwinds from a continuously changing and challenging environment,' Minister Rajah began. Singapore faced 'two winds of change – the environmental headwinds, and political headwinds' and called for 'decisive and concerted action to mitigate these risks.' The property catastrophe re/insurance industry, 'by combining its risk financing capacity, with its risk mitigation capabilities, can play a huge role in managing these risks,' the minister argued. Minister Rajah in turn outlined several interconnected re/insurance market development initiatives underway with Singapore state support, including the recent launch of a new insurance-linked securities trading market with the capacity to finance catastrophe risk for Asia.

The minister's remarks reflect the ways in which coalitions of states, multilateral organisations like the World Bank, and financial institutions increasingly turn to the property catastrophe re/insurance sector to manage the entwined ecological, political, and economic uncertainties of climate change. Re/insurers constitute a lucrative, multi-billion dollar risk financing system, one which offsets catastrophe losses across a wide range of geographies. Global reinsurers, or insurers for insurers, promised $625 billion of protection capital to their clients in 2019, for example (Aon Benfield 2020). Re/insurers have also emerged as prominent proponents for (and investors) in a range of climate risk finance experiments. This can be seen in the roll-out of multilateral disaster risk pools and other insurance products in support of several sustainable development and humanitarian agendas (Grove 2012; Johnson 2021), or through the extension of re/insurance instruments and models services to help non-insurance financial institutions like real estate

DOI: 10.4324/9781003157571-15

asset managers to govern their physical climate risk exposure (Taylor and Aalbers 2022). Re/insurers curate and perform epistemologies of risk, build institutional capacities and tools to manage such risk, and profit from a wide array of risk finance and advisory services (Taylor & Weinkle 2020). They play multifaceted roles in the assembly and expansion of regimes of financialised disaster risk governance (Grove 2012), and by extension mediate the moral economies of climate change in powerful ways (Elliott 2021).

Despite this global(-ising) influence, re/insurance markets are geographically uneven, contingent, and provisional (Johnson 2013, 2014; Taylor & Weinkle 2020; Booth 2021). The market's contemporary capital flows and expertise largely remain confined to regions with risks that are sufficiently profitable to lure capital, actuarially well-defined enough to be priced with confidence, and where other conditions (like favourable state regulation) enable and ensure market access. As re/insurers seek to construct new risk capital markets across emergent geographical frontiers, they must also contend with recurrent frictions and dislocations within existing market territories, ranging from debates over insurance affordability, to post-disaster crises of insolvency and market abandonment. Growing anxieties about the 'uninsurability' of a number of well-established underwriting domains – ranging from fossil fuel infrastructure to US coastal real estate – are contemporary examples of the existential headwinds facing the sector.

How do we reconcile the globalising yet provincial, universalising but contingent character of these markets at this crucial juncture, as state and capital alike seek to manage a world of unruly climate risk using re/insurance models, methods, and capital instruments? This chapter responds to these tensions by examining how re/insurance markets evolve through the relational interplays within and between key urban geographies in the hurricane wind risk trade. Contemporary efforts to construct an insurance-linked securities (ILS) and catastrophe finance hub in Singapore provide the touchstone for this essay. By transforming insured risks into an investment asset class, ILS instruments are widely seen to be key to securing the capacity of re/insurance markets to finance new and expanding horizons of catastrophe risk (Johnson 2013, 2014; Taylor 2020)[2]. Singapore's first full-fledged, SEC Rule 144A-compliant ILS was issued on behalf of a Florida insurer, backing the hurricane wind risk exposure within thousands of residential property insurance policies. Florida relies on ILS and other forms of re/insurance capital to finance its hurricane-exposed, real estate-driven political economy. Well-established Florida re/insurance risk capital flows have provided an ideal test case to demonstrate Singapore's competence as an ILS hub, as part of a larger play to capture a growing share of the Asian re/insurance business, and in turn to secure the city-state's advantageous, if precarious position as an international financial centre.

The chapter develops the Florida-Singapore ILS case to make two contributions to a growing body of critical insurance studies (Booth 2021). First, the case deepens our understanding the relational ways in which specific urban geographies are central to the assembly and extension of re/insurance markets. Florida and Singapore play crucial roles in constituting catastrophe re/insurance markets (as a source of insured wind risk, or as a centre for brokering such risk, respectively). Yet as the chapter explores, so too do actors operating in each context seek to cultivate re/insurance in order to hedge against broader, yet distinctive political and economic 'headwinds,' ranging from an uncertain future for risky coastal real estate in Florida, to anxieties about regional competition and stability in Singapore. The chapter underscores how these interwoven, if asymmetrical set of relations shape and extend re/insurance as a powerful modality for governing climate uncertainties. Second, the chapter aims to encourage further relational analysis of the intra- and inter-geographical dynamics which shape the scope and significance of re/insurance geographies. Three analytical focal points – (i) circulations of tools and techniques, (ii) risk capital flows, and (iii) shifting state engagement – are proposed and explored in conversation with the case to advance relational re/insurance analysis.

Following this introduction, the chapter charts the evolution of ILS within and between Florida and Singapore. In turn, the chapter draws on insights from the case to develop the three aforementioned focal points for the relational study of re/insurance market change. The chapter draws on fieldwork conducted in Florida and Singapore between 2016 and 2019, including elite interviews with re/insurance executives and other market stakeholders, public policy and financial statement analyses, and in-person and virtual participation at major industry conferences and events, including RMS Exceedance (2016), the Singapore Reinsurance Conference (2019), and Artemis ILS Asia Conference (2020).

Florida: Underwriting urban fortunes

Re/insurers often characterise Florida as the 'peakest' of 'peak peril' property catastrophe underwriting, due to the exceptional concentration of insured hurricane wind exposure in the state. Swiss Re estimates that a single major-category hurricane[3] landfall in Miami could generate insured losses of up to $180 billion, and economic damages far greater, representing losses of 'a magnitude not yet observed' by the industry, for example (Schwartz & Linkin 2017). To manage this risk, re/insurers collect substantial volumes of policyholder premiums from millions of Florida policyholders every year. In 2018, Florida retail insurers directly collected more than $10 billion in annual premiums, underwriting over $2 trillion of statewide residential exposure (Florida Office of Insurance Regulation. n.d.). Florida insurers cede a large proportion of this premium to dozens of catastrophe reinsurers in Bermuda, the United Kingdom, Germany, and beyond, which agree to finance a share of the insurer's catastrophe risk exposure. Taylor (2020) finds that an important subset of Florida residential insurers spent just over half of every consumer premium dollar earned on reinsurance in 2015, for example.

The size of the residential insurance business in Florida is a function of how and where the state has urbanised in the post-World War II era. Real estate became a core driver of the state's economy, the profitability of which has generally exceeded longstanding concerns about the state's fragile, hurricane- and flood-prone coastal geography (Audirac et al. 1990; Catlin 1997). Over this horizon, local 'growth machines' took root, which became structurally reliant on sustained development to generate property tax receipts, real estate-related jobs, and related service-sector employment (Taylor 2020). At the same time, federal government programs and regulations subsidised post-war urbanisation through growth-inducing and environmental risk-reducing infrastructures (as in the federal highway system, or Army Corps of Engineers projects), through the widespread (but not universally accessible) expansion of mortgage markets, and other spatially redistributive practices, which disproportionately favoured Florida and other 'sunbelt' states (Bernard & Rice 1983).

The substantial human, ecological, and economic devastation wrought by hurricanes, including Andrew (1992), those of the 2004 and 2005 Atlantic hurricane seasons, and Irma (2017), have focused attention on Florida's environmental precarity more generally, and on the fragility of the real estate-driven political economy built thereupon specifically. Re/insurance became the de facto 'fix' for Florida's risky real estate dilemma for two closely related reasons according to Taylor (2020). First, federal government housing finance regulations institutionalised the use of multi-peril property insurance within the residential property finance market, creating a structural role for insurance within the US housing finance system. Second, decades of pro-growth urban governance in Florida saw rates of building in catastrophe-prone areas far exceed the use of land use controls, building codes, and infrastructure investment to curtail the rise of catastrophe exposure in the built environment. Not only did Florida become economically dependent on ecologically fraught patterns of development, it did so in ways which relied on re/insurers to finance property catastrophe risk, in the absence of meaningfully integrated and comprehensive urban environmental risk management. An executive at a major engineering, construction, and design firm in Florida reflected this sentiment in an interview about contemporary resilience planning efforts in the state:

> So far, there hasn't been much discussion about the real players in this: the re/insurance industry. [...] Eventually, you can have all the politics and all the plans you want, but this private sector will eventually have to come to the table. If they come to the marketplace too quickly, they'll destroy value in the market, which is not the value of a resilience program.
>
> (Interview 2018-A)

The Florida market has presented industry-defining challenges and opportunities to global re/insurers and state public policymakers

(Medders et al. 2013; Taylor 2020). These pivot around the complexities of maintaining an underwriting regime that is reliable and profitable, while also sufficiently affordable to property market consumers – and therefore 'resilient' enough to sustain the state's real estate-based political economy (Taylor & Weinkle 2020). Questions of how to actuarially model and price potential catastrophe hurricane wind losses, to capitalise this risk and transfer it from insurers to reinsurers and broader capital market investors, and to appropriately regulate such practices to serve broader societal goals have informed debates and innovations in the re/insurance sector in the three decades since Hurricane Andrew's landfall near Miami (Medders et al. 2013; Taylor & Weinkle 2020).

ILS products were introduced and eventually widely adopted in Florida and beyond in response to such questions. One senior re/insurance investor estimated that half of all ILS are exposed to Florida hurricane risk, for example (Seo 2015). Between March 2017 and June 2020, Florida residential insurers channelled millions of dollars in policyholder premiums to raise $6.2 billion of reinsurance protection through 33 public ILS issuances (Table 11.1). This count does not include ILS transactions sponsored by reinsurers transferring their Florida risk exposure nor does it include privately placed collateralised reinsurance, and therefore represents only a portion of ILS-related activity in the state.

Table 11.1 Florida insurer direct ILS issuance (March 2017–June 2020).

Florida insurer	*ILS issuances*	*ILS cover (cumulative, millions)*
Citizens Property Insurance Corporation	4	$685.0
Heritage Property and Casualty Insurance Co.	2	$160.0
Safepoint Insurance Company	2	$240.0
Southern Oak Insurance Company	2	$99.2
Avatar Property and Casualty Insurance Company	2	$165.0
Castle Key Insurance & Castle Key Indemnity	2	$400.0
USAA	8	$2,040.0
Security First Insurance Company	2	$275.0
American Integrity Insurance Company via Hannover Rück SE (Germany)	4	$489.0
American Strategic Insurance Group	1	$200.0
Nationwide Mutual	3	$1,315.0
State National Insurance Co via Markel Bermuda Ltd	1	$100.0

Notes: Bonds with exclusive Florida exposure and mixed US multiple peril exposure (including named storm) were included. While many of these firms are mostly or entirely specialised in Florida underwriting (e.g. Citizens), some (e.g. USAA, Nationwide) are large national insurers with diversified portfolios and catastrophe risk exposures.

Source: Artemis.bm Deal Database.

Decades-long experiments in capitalising Florida insurers through ILS and other reinsurance instruments have made Sunshine State hurricane wind risk a well-established asset class, for which there exists a significant depth of industry, investor, and regulator familiarity. The geographical specificity of Florida's crisis-prone mode of urbanisation, and the industry's role and response in developing instruments like ILS to hedge against such crisis, helps to foreground how and why Singapore's first full-fledged catastrophe bond issuance should come to be underpinned by Florida residential hurricane wind risk.

Singapore: Regionalising risk capital

Surveying Singapore's glass-and-steel financial district skyline from a high-rise conference room, a reinsurance executive recounted the collaborative efforts of state and industry actors to construct a local ILS trading centre in the city-state, which raised just short of $1 billion of catastrophe coverage between the first such issuance in 2019 and mid-2020 (Table 11.2). The executive detailed how, over the course of years of dialogue and through the roll-out of market-making infrastructures, the foundation was laid for

Table 11.2 Singapore-based ILS transactions, May 2019–June 2020.

Bond issuance	Cover (millions)	Peril(s)	Sponsor
First Coast Re II Pte. Ltd. (Series 2019-1)	$100	Florida-named storm & severe thunderstorm	Security First Insurance Co (Florida)
Manatee Re III Pte. Ltd. (Series 2019-1)	$40	US-named storm & severe thunderstorm (Florida, Louisiana, New Jersey & Texas)	Safepoint Insurance Co (Florida)
Integrity Re II Pte. Ltd. (Series 2020-1)	$150	Florida-named storm	American Integrity Insurance Co (Florida) via Hannover Rück SE (Germany)
Akibare Re Pte. Ltd. (Series 2020-1)	$100	Japan typhoon & flood	Mitsui Sumitomo Insurance Co (Japan)
Catahoula Re Pte. Ltd. (Series 2020-1)	$60	Louisiana-named storm & severe thunderstorm	Louisiana Citizens Property Insurance Corporation (Louisiana)
Casablanca Re Pte. Ltd. (Series 2020-1)	$65	Florida-named storm	Avatar Property and Casualty Insurance Co (Florida)
Alamo Re II Pte. Ltd. (Series 2020-1)	$400	Texas-named storms & severe thunderstorms	Texas Windstorm Insurance Association (Texas)

Source: Artemis.bm Deal Directory.

the first ILS issuance in Singapore (Interview 2019-A). Legal frameworks were retooled to accommodate a new offshore special purpose reinsurance institutional structure, while regulatory structures were revised to make oversight and compliance streamlined and cost-effective for issuers and investors. An array of specialist professional services providers (brokers, lawyers, actuaries) had to be recruited or trained. A new state-funded grant scheme was launched to attract issuers, and to offset the 'frictional' costs of issuing a transaction in unfamiliar territory. Asian regional investment managers had to be thoughtfully introduced to this new asset class and courted for future deals.

First Coast Re and the transactions which followed represented a crucial proof of concept, a symbolic and strategic 'practical accomplishment' (Fields 2018) within a larger play to cultivate an expanded role of Singapore as a broker within global catastrophe risk finance circuits. First, the growth of catastrophe re/insurance and risk finance-related services is seen by aligned actors as a means to extend and secure the scope and scale of Singapore's international finance centre. Financial services comprise an increasingly important driver of Singapore's unique state-capitalist economic development model (Olds & Yeung 2004; Chua 2017), growing from 4.6% of Singapore's GDP in 1965, to 12.3% by 2016 (Lai 2018, p. 154). Re/insurance is one of several financial subsectors which have registered state investment in recent years (Lai & Samers 2017; Lai 2018; Dodge 2020). In line with this strategy, many activities like manufacturing have been gradually relocated to neighbouring countries (yet often remain under the control of Singaporean enterprises), in favour of the growth of higher-wage advanced producer services, including finance (Chua 2017; Lai 2018). By 2020, Singapore's financial sector employed more than 170,000 workers, yielding 13.3% of the GDP despite accounting for only 4.5% of the workforce (Monetary Authority of Singapore 2020). Financial institutions also comprise a significant source of demand for 'Grade A' office space, the sustained re/development of which is an important driver and feature of Singapore's unique property-driven state capitalist model (Haila 2016).

Absent the need to finance large volumes of domestic property catastrophe exposure (as in Florida), Singapore-based re/insurance institutions specialise in brokering risks across Asia and Australia. In 2018, Singapore's offshore re/insurance hub wrote $12.8 billion in gross premiums, of which nearly 60% was in property lines (Monetary Authority of Singapore n.d.). The largest sources of premium were China (34.9%, exclusive of Hong Kong), Japan (13.7%), Australia (10.8%), and Thailand (8.3%) (ibid). Nevertheless, the extent of re/insurance activity in Singapore remains modest in comparison to larger reinsurance hubs like London, Bermuda, and Zurich. In 2017, London captured $110 billion of premium, or roughly ten times that of Singapore (London Market Group 2020, p. 2).

Long-term visions for Singapore's re/insurance sector therefore tend to focus on exploiting the city'-state's access to Asian risks and capital. In conference presentations and interviews, re/insurers argue that the combination of growing regional climate risk and high economic growth has yet to be matched by the rate of property re/insurance market take-up, suggesting an array of underwriting opportunities on the horizon (Monetary Authority of Singapore 2019; 2020). In this context, efforts to expand the Singapore centre may be seen as one of several broader state, multi-lateral, and financial market institutional manoeuvres to extend re/insurance underwriting across Asia. At the same time, the ongoing rise of an Asian investor class is seen to represent a vast pool of regional capital that could be deployed as re/insurance capacity through instruments like ILS. Singapore's re/insurance proponents have sought to meld these elements by expanding the offering of products and services available, including ILS issuance and investment management (Interview 2019-B). One Singapore-based reinsurance executive hypothesised that the ultimate aim of Singapore's market-makers was not to rival London or Bermuda in scale, but instead to provide high value-added financial solutions for specialist regional underwriting and investment needs. The executive thus likened the Singapore ILS strategy to a private jet, one able to seat only a handful of precious customers, and with each issuance representing one such seat on the jet (Interview 2019-C).

The development of ILS markets also arguably advances a second political economic agenda for the city-state, one rooted in securing regional stability through catastrophe risk finance. Singapore's economic development strategy has long been informed by recognition of the city-state's precarious geographical position. While this may be acutely true in the case of finance – neighbouring Hong Kong has also set out to develop an offshore ILS hub (Lim et al. 2020) – it also broadly applies to the future of the resource-constrained island nation. Singapore relies on neighbouring nations for many essential inputs, including water, food, and labour. The expansion of Singapore's advanced producer services economy also hinges on the continued political and economic stability of neighbouring countries, given that a significant share of regional economic activities and investments are underwritten by Singaporean enterprises or coordinated by Singapore-based financial institutions (Olds & Yeung 2004). In this context, the economic regionalisation goals of Singapore are intimately linked with anxieties about state security (Lee 2001; compare with Grove 2012).

The expansion of Singapore's re/insurance centre aims to sustain these transnational and intra-regional ties in the face of catastrophic disruption due to climate risk. Parallel to efforts to draw ILS issuance and investment through Singapore, the state and re/insurance institutions have co-sponsored research on regional catastrophe risk modelling at Singapore universities (Interview 2020-D). At the same time, the Singapore government is a host of (and investor in) the Southeast Asia Disaster Risk Insurance Facility (SEADRIF), a World Bank-driven sovereign risk pool which aims to raise risk capital on behalf of ASEAN nations without established retail insurance markets. The opportunistic decision to grow Singapore's catastrophe re/insurance capacities by leveraging ILS issuance drawn from Florida hurricane wind risk and beyond can therefore also be understood as a project which seeks to secure Southeast Asia's economic trajectory against catastrophic disruptions – and Singapore's advantageous if fragile position therein.

Discussion: Mapping headwinds on the horizon

Florida and Singapore constitute, and in many ways are constituted through, urban geographies of hurricane wind re/insurance exchange. The Florida-Singapore ILS case illuminates how re/insurance markets emerge and expand through disparate yet interwoven ecological, political, and economic dynamics within and between geographies. While Florida continues to serve as an industry-defining source of insured wind risk to be marketised or otherwise managed through re/insurance, Singapore has emerged as a key centre for brokering such risk for new investors, as part of a broader strategy of financialised regional catastrophe risk management (see Grove 2012). Actors operating within and between each context cultivate re/insurance in response to a plurality of 'headwinds' on the horizon, ranging from anxieties about the insurability of growing property catastrophe risk exposure and the search for a safe haven for collateral-seeking capital, to the need to pre-emptively secure particular regional and sectoral relations against destabilisation-by-disaster. While questions about the long-term insurability of particular assets, places, or perils are among the existential headwinds facing re/insurers, so too does this sector remain a powerful force when it comes to defining and managing unruly climate uncertainties through finance.

Continued relational analysis of re/insurance markets is vital to our understanding of the broader geographies of climate governance. As a gesture towards this open-ended project, this concluding discussion proposes three analytical focal points for such a relational approach, in dialogue with the Florida-Singapore ILS case. First, the case reiterates how *circulations of tools and techniques* work to secure or extend re/insurance geographies. Insurance-linked securities and the catastrophe risk models used to marketise risks therein, play a constitutive role in the Florida-Singapore ILS case, yet their origin, adoption, and adaptation by the industry remain rooted in specific geographical sites and practices (Jarzabkowski et al. 2015; Taylor & Weinkle 2020). Although catastrophe risk models are integral to contemporary actuarial practice, their initial take-up was closely linked with efforts to address the particular challenges of predicting and pricing low probability, high-value catastrophic loss events, and Florida hurricanes in particular. Models must be continually adapted to enable their deployment across new regions and perils, coevolving with industry investments in data-capture and synthesis to tap (or induce) new market demand, advancements in scientific understandings of particular risks, and changing non-financial stakeholder perceptions of the value and usefulness of risk models. In Singapore, state and re/insurance industry figures are co-sponsors of scientific research which seeks to translate risk science and existing actuarial capacities to the valuation of perils in Southeast Asia. A history of ILS and its uses could be drawn along similar lines, as instruments have been developed in relation to underwriting capitalisation issues in troublesome submarkets like Florida, and in turn retooled to fund a wide variety of insured perils, including those beyond the horizons of natural disaster risk. Continued attention to the contexts in which actuarial technologies are assembled, adapted, or abandoned can provide a fruitful line of analysis for understanding re/insurance market transformation.

The case also underscores how the analysis of *risk capital flows* helps to reveal where, why, and how re/insurers shape urban-material ties and 'socialities' (Christophers et al. 2020) within and between places and actors. First Coast Re and other ILS organise geographically disparate risks and capital flows to serve multiple if contingent political and economic goals. ILS provides a means through which the industry markets insured risk as an asset class accessible to investors seeking new horizons of risk and return not correlated with the broader economy. Not only do alternative reinsurance products open up a new horizon of accumulation for investors, they also enable financial services firms to capture economic value from risk management services, like brokering and modelling, the activities and capital flows of which undergird international financial centres such as Singapore and London. At the same time, ILS represents a promise to pay to insurers and their policyholders, the confidence in which underwrites a broader range of financial and non-financial activities. From employment to public sector fiscal capacity, Florida's real estate-driven political economy depends on access to re/insurance risk capital. Following risk capital flows through market geographies reveals crucial points of tangency and logics of interdependency between insurance and other political-economic dynamics, which can illuminate the structural importance as well as potential limits of re/insurance within specific geographies. For example, Taylor's (2020) analysis of ILS flows in Florida illuminates the state's expensive reliance on risk capital markets, opening up questions about the array of (extra-) economic values selectively re/produced through re/insurance (Elliott 2021) and the variegated links between finance, property catastrophe exposure, and urban restructuring (Taylor & Aalbers 2022).

As scholars of financialisation have long argued, attention to *shifting state engagement* also sheds light on the geographical presences, absences, and varying public purposes of re/insurance markets. Mainstream insurance scholars and market advocates have at times conceptualised re/insurance using a false binary between the state and market (Taylor & Weinkle 2020), which obscures our understanding of the role of public regulations, investments, and other activities in shaping – or even creating and destroying – the contours of re/insurance markets. The Florida and Singapore contexts reveal multiple examples of entrepreneurial forms of state intervention in support of the construction of local re/insurance markets, which underpin distinctive urban political-economic projects. Florida's Citizens Property Insurance Corporation has been among the most prolific catastrophe bond issuers in recent years (Table 11.1). Public insurers in Louisiana and Texas are also among early issuers in Singapore (Table 11.2). Beyond issuing ILS, Florida state agencies invest in the market, and shape private insurer demand for reinsurance through regulations, subsidies, and guarantees, as part of a broader play to secure its housing- and property development-driven political economy (Taylor 2020). Similarly, the Singapore government maintains a grant scheme for ILS market entrants, which offsets the costs of preparing a new issuance. The case also highlights indirect forms of state-firm collaboration and intervention, like public-private partnerships which sponsor risk-related scientific research in Singapore, or state-sanctioned real estate growth strategies which reproduce a structural market for re/insurance in Florida. While states actively seek to extend re/insurance markets to further particular policy outcomes, it is important to recognise that inherited public policies, enduring path dependencies, and evolving state capacities can also normalise or otherwise enable specific re/insurance market patterns and logics. How, then, do these patterns of statecraft shape how, where, and when urban geographies come to be entangled with re/insurance? And how might they be repurposed to confront emerging dilemmas raised by climate disruptions?

These three focal points – circulations of tools and techniques, risk capital flows, and shifting state engagement – seek to populate a critical imagination for how we might continue to trace the evolving project of re/insurance within and between urban geographies, in the face of complex headwinds on the horizon. Insofar as these risk underwriting institutions increasingly govern diverse configurations of ecological, economic, and political contingencies against climate uncertainties through the universalising rubrics and rationalities of finance, so too must they be understood to be simultaneously provincial and cosmopolitan, contingent yet interdependent, powerful yet malleable.

Acknowledgements

The author wishes to thank Ricardo Cardoso and Jane M. Jacobs for the opportunity to conduct research in Singapore as a Visiting Scholar at Yale-NUS College in Autumn 2019, to Rachel Bok, Liliana Gil, and Andrew Moon for lively debates in and about Singapore, and re/insurance industry members for interviews and access to their spaces of practice. Thanks is also extended to the organisers and participants of the 'Climate, Society, and Insurance: Capacities and Limitations' International Academic Forum at the University of Tasmania and to Manuel Aalbers and members of the Real Estate/Financial Complex Research Group at the University of Leuven for their feedback on earlier drafts of this work. This research was funded by a H2020 Marie Skłodowska-Curie Actions Fellowship (ID: 799711).

Notes

1 Re/insurance is shorthand for the insurance and reinsurance market. Reinsurance is a form of insurance for insurers.
2 While ILS refers to a specific subset of insurance-based financial instruments, it also signals a larger universe of 'alternative capital' arrangements within the re/insurance industry.
3 Hurricanes in the Atlantic Basin are categorised by wind speed along the Saffir-Simpson Scale: Category 1 (74–95 mph, 119–153 km/h), Category 2 (96–110 mph, 154–177 km/h), Category 3 (111–129 mph, 178–208 km/h), Category 4 (130–156 mph, 209–251 km/h), Category 5 (≥157 mph, ≥252 km/h). Categories 3-5 storms are classified as 'major.'

References

Aon Benfield 2020, 'ILS annual report 2020', *Alternative Capital: Growth Potential and Resilience*, Chicago: Aon Benfield.

Audirac, I, Shermyen, A & Smith, M 1990, 'Ideal urban form and visions of the good life: Florida's growth management dilemma', *Journal of the American Planning Association*, vol. 56, no. 4, pp. 470–482.

Bernard, R & Rice, B 1983, *Sunbelt Cities: Politics and Growth since World War II*, Austin: University of Texas Press.

Booth, K 2021, 'Insurance, and the prospects of insurability', In Knox-Hayes, J & Wojcik, D (eds), *The Routledge Handbook of Financial Geography*, London: Routledge.

Catlin, R 1997, *Land Use Planning, Environmental Protection, and Growth Management: The Florida Experience*, Chelsea, MI: Ann Arbor Press.

Christophers, B, Bigger, P & Johnson, L 2020, 'Stretching scales? Risk and sociality in climate finance', *Environment and Planning A*, vol. 52, no. 1, pp. 88–110.

Chua, B.H 2017, *Liberalism Disavowed: Communitarianism and State Capitalism in Singapore*, Singapore: NUS Press.

Dodge, A 2020, The Singaporean natural gas hub: Reassembling global production networks and markets in Asia, *Journal of Economic Geography*, vol. 20, no. 5, pp. 1241–1262.

Elliott, R 2021, *Underwater: Loss, Flood Insurance, and the Moral Economy of Climate Change in the United States*, New York: Columbia University Press.

Fields, D 2018, 'Constructing a new asset class: Property-led financial accumulation after the crisis', *Economic Geography*, vol. 94, no. 2, pp. 118–140.

Florida Office of Insurance Regulation n.d., *Quarterly Supplemental Report – Market Share Report, All Insurers, All Residential Lines, Q4 2018*, <www.apps.fldfs.com/QSRNG/Reports/ReportCriteriaWizard.aspx>.

Grove, K 2012, 'Pre-empting the next disaster: Catastrophe insurance and the financialization of disaster management', *Security Dialogue*, vol. 43, no. 2, pp. 139–155.

Haila, A 2016, *Urban Land Rent: Singapore as a Property State*, Chichester, UK: John Wiley & Sons Ltd.

Jarzabkowski, P, Bednarek, R & Spee, P 2015, *Making a Market for Acts of God: The Practice of Risk Trading in the Global Reinsurance Industry*, Oxford: Oxford University Press

Johnson, L, 2013, 'Catastrophe bonds and financial risk: Securing capital and rule through contingency', *Geoforum*, vol. 45, pp. 30–40.

Johnson, L 2014, 'Geographies of securitised catastrophe risk and the implications of climate change', *Economic Geography*, vol. 90, no. 2, pp. 155–185.

Johnson, L 2021, 'Rescaling index insurance for climate and development in Africa', *Economy & Society*, vol. 50, no. 2, pp. 248–274.

Lai, K 2018, Singapore: Connecting Asian markets with global finance In Cassis, Y & Wójcik, D (eds.) *International Financial Centres: After the Global Financial Crisis and Brexit*, Oxford: Oxford University Press.

Lai, K & Samers, M 2017, 'Conceptualizing Islamic banking and finance: A comparison of its development and governance in Malaysia and Singapore', *The Pacific Review*, vol. 30, no. 3, pp. 405–424.

Lee, W 2001, 'Economic restructuring in Singapore: A reflection on regional security in Southeast Asia', *Asian Affairs: An American Review*, vol. 27, no. 4, pp. 211–221.

Lim, T.L, Spitzer, R & Yi, J.W.U 2020, Hong Kong on the Verge of Launching an ILS Regime, *Mayer Brown*, May 11.

London Market Group 2020, *London Matters 2020 Annual Report*, London: London Market Group.

Medders, L, Nyce, C & Karl, J 2013, 'Market implications of public policy interventions: The case of Florida's property insurance market', *Risk Management and Insurance Review*, vol. 17, no. 2, pp. 183–214.

Monetary Authority of Singapore, n.d. *2018 Insurance Data, Table AG 10: Gross Premiums of Offshore Insurance Fund Business by Line*, Retrieved online, viewed 18 December 2020, <www.mas.gov.sg/statistics/insurance-statistics/annual-statistics/insurance-statistics-2018>.

Monetary Authority of Singapore 2019, *Keynote Address by Ms Indranee Rajah, Minister in the Prime Minister's Office and Second Minister for Finance and Education at the 16th Singapore International Reinsurance Conference* (SIRC), 30 October 2019, viewed 18 December 2020, <www.mas.gov.sg/news/speeches/2019/keynote-address-by-ms-indranee-rajah>.

Monetary Authority of Singapore 2020, *Gearing Up for New and Evolving Jobs in Financial Services*, Remarks by Mr Ravi Menon, Managing Director, Monetary Authority of Singapore, at 'Growing Timber' MAS-IBF Webinar Series on 26 November 2020, viewed 18 December 2020, <www.mas.gov.sg/news/speeches/2020/gearing-up-for-new-and-evolving-jobs-in-financial-services>.

Olds, K & Yeung, H 2004, 'Pathways to global city formation: A view from the developmental city-state of Singapore', *Review of International Political Economy*, vol. 11, no. 3, pp. 489–521.

Schwartz, M & Linkin, M 2017, *Hurricane Andrew: The 20 Miles that Saved Miami, Armonk, USA*, Swiss Reinsurance Company.

Seo, J 2015, *Statement before the US Department of Treasury*, Federal Advisory Committee on Insurance Hearing, 4 November, <www.yorkcast.com/treasury/events/2015/11/04/faci/part-2>.

Taylor, Z 2020, 'The real estate risk fix: Insurance-linked securitization in the Florida Metropolis', *Environment and Planning A*, vol. 52, no. 6, pp. 1131–1149.

Taylor, Z & Aalbers, M. 2022, 'Climate gentrification: Risk, rent and restructuring in Greater Miami', *Annals of the American Association of Geographers*, DOI: 10.1080/24694452.2021.2000358

Taylor, Z & Weinkle, J 2020, 'The riskscapes of re/insurance', *Cambridge Journal of Regions, Economy and Society*, vol. 13, no. 2, pp. 405–422.

12 Emotions and under-insurance

Exploring reflexivity and relations with the insurance industry

Nick Osbaldiston

Introduction

In the social sciences, specifically sociology, the question of how we understand, perceive, and act on knowledge is well theorised especially through the concept of reflexivity. As societies move away from the traditional modes of knowledge, living, and structure (e.g. religion and mythology), we are opened to a world flooded with scientific knowledge and understandings of risk (Beck 1992). Expertise, in this type of society, is paramount to interpreting what is risky now, and how to go about facilitating adaptation or avoidance of the dangers of modern life. Furthermore, because our life-choices are now far more open than prior, we are consistently bound to a life-project where specifically identities are played with, adopted, and reconfigured daily (Giddens 1990).

Theories of risk and reflexivity have in recent times included emotions in their schemas and frameworks (Burkitt 2012; Holmes 2010, 2015; Lowenstein et al. 2001; Sjoberg 2007; Slovic et al. 2005). Acknowledged amongst this growing scholarship is the way emotions inform and accompany decision-making and risk-perception in both large-scale events and in everyday lives (Hogarth et al. 2011). The rational system that is used to process and analyse does not necessarily triumph over the more affective system, but rather, can be coloured by it. The emotional dimensions of memories, images, stories, and cultural codes influence the processing and analysis of information and what this information means to us personally (Slovic et al. 2005).

In this chapter, I seek to revisit these conceptualisations to understand how people process, interpret, and understand their relationship with insurance and risk. Using empirical examples from a study conducted in cyclone-prone Far North Queensland, Australia, I aim to show that reflexivity is not simply a matter of cognition. Rather, people come to understand insurers and the dangers of losing their property through schemas of logic *and* emotions.

This research adds to a growing emphasis on emotions/affect in decision-making especially within financial practices (Rick & Lowenstein 2008). Specifically, the ways emotions play a role in the lead-up to purchasing consumer items and how the expected emotions can confirm or challenge initial

DOI: 10.4324/9781003157571-16

purchase feelings is explored here (Rick & Lowenstein 2008; Lowenstein et al. 2001). The significance of this chapter, however, lies in its focus on insurance – how the insurance can be deemed risky and how emotional responses to insurance emerge in relation with others. This is an area of research that has not received significant attention in sociology.

Furthermore, this research reflects the importance of renters who are often overlooked in insurance debates. Booth and Kendal (2020) suggest that renters are deemed to be at less risk by institutions such as local governments due to a lack of property ownership and assumed low asset base. However, they argue that 'without adequate insurance,' renters face potentially face 'homelessness or having no option but to live in damaged property' (Booth & Kendal 2020, p. 742). As Osbaldiston, McShane and Oleszek (2019) have shown, there is also evidence of increasing non-insurance amongst renters likely due to costs, self-efficacy and confidence.

Risk, reflexivity, trust, and emotions

At a general level, reflexivity is the emergence of a state of cognition and reflection in our modern times (Giddens 1990; Beck 1992). Giddens (1990, p. 38) describes this condition as consisting;

> in the fact that social practices are constantly examined and reformed in the light of incoming information about those very practices, thus constitutively altering their character.

The reflexive mind then, for Giddens (1990), is one which is consistently drawing on new information to monitor our sense of self and identity, and then adapt accordingly. As traditional society, which enforced certain options on us (e.g. family, religion), begins to fade, our worlds are opened to interpretation and importantly, different actors who claim expertise. In place of traditional gatekeepers of information, are a proliferation of experts in areas as diverse as school counselling through to life coaches.

However, with the proliferation of uncertainty, there arises a questioning of the scientific expertise that previously made things 'certain.' Specifically, events like Chernobyl, especially for Beck (1992), have made society sensitive to the risks that modernity has produced. Subsequently, institutions such as the state are no longer trusted to have all the answers and people can draw upon other expertise to make their judgements about what to do in response to diverse risks. This processing of different opinions and counter-opinions from 'so called' expertise is what defines reflexivity. *Who* to invest your trust into, *how* we derive this decision and *what* form that trust relationship takes, need to be studied in the social sciences.

For Beck (1992) and Giddens (1990) though, this work is mostly cognitive as we take on information from expertise and make decisions on who to invest our trust in. As noted above, our understandings of risk perception have shifted to include thinking through emotions as an additional variable. 'Probability judgements,' writes Slovic et al. (2005, p. 36), can be influenced by, if not determined through, 'readily available affective' impressions which require far less 'complex or mental resources.' This consideration is one of the criticisms that is levelled at Giddens (1990) and Beck's (1992) interpretation of risk and reflexivity. Specifically, the argument suggests that 'lay actors,' while 'rational,' experience their worlds through emotions, bodies, and symbols (Mythen 2005, p. 144). Frames of 'risk reference' need to be expanded in our analysis as individuals will select from a wide range of factors in their decision-making than is often presented in theories of risk and reflexivity (Mythen 2005, p. 144).

In recent times, sociology has engaged with this further by rethinking reflexivity in relation to how people process feelings. Holmes (2010, 2015), for instance, argues that our era is defined by uncertainty requiring us to place some trust in expertise. She suggests that 'trust [...] is often necessarily based largely on emotions, on feelings about things and activities, or an aesthetic – a liking for a person, persons or thing' (Holmes 2010, p. 143). In other words, we do not simply rely on cognition to interpret our social worlds, and importantly, on who we place our trust in. Rather;

> reflexivity is not simply a rational calculation of the amount of satisfaction an aspect or way of life brings, but it is infused with feelings about how it fits (or does not) with others and what they think, feel and do. *Reflexivity is emotional* and comparative and relies on interpreting emotions.
>
> (Holmes 2010, p. 148, italics added)

Importantly, these emotions do not emerge in isolation. They are derived through our interactions and relations with others.

Emotions are done in interaction with others; they involve bodies, thought, talk, and action. Feelings make embodied social selves and selves and lives are made within the social constraints of place and time. It is crucial to attempt to better understand these emotional reflexive practices within a sociological context (Holmes 2010, p. 149).

Not only do our relations with others manifest certain types of emotions that then colour our decision-making, but our positions in the social world also inform certain emotional responses to different events. Burkitt (2012, p. 466) makes this point when he supposes that if we were caught 'being dishonest by someone we highly regard as a person of integrity, we will feel doubly ashamed.' The judgement of others is as important as the actual event that precedes it.

Thus, context matters. In understanding insurance through emotional reflexivity, the emotions one feels towards insurers, social peers, the geography of the area, and past experiences of natural disasters perhaps all have a role in defining how one engages with insurance. While for some, insurance is not a matter of trust per se, but of a social norm defined by macroeconomic policies and practices (Tranter & Booth 2019), others who have more agency to choose to insure or not (such as renters) tend to reflect on their relationship to insurers, climates and geographies with more emotional reflexivity. It is this I seek to explore in the rest of the chapter.

Methodology

The empiric in this chapter is based on research conducted in 2017 and again in 2020 exploring the experiences and perceptions of householders towards risks in the tropical North Queensland city of Cairns. With a population of over 150,000 and a geography that lay adjacent to the Coral Sea exposing the place to cyclone risks, Cairns has in recent times suffered dramatic increases in insurance costs (ACCC 2019). This rise in premiums, which is suffered across Northern Australia generally, has led to some evidence of underinsurance and non-insurance (ACCC 2019; Osbaldiston, McShane & Oleszek 2019). The latter is defined here as simply those households which have elected to not purchase a policy that covers risk to property and/or possessions. The former refers to those households which have insurance cover which is inadequate to fully replace/rebuild property and/or possessions (Booth & Harwood 2016).

To investigate how people in Cairns engage with insurance, forty people were interviewed about their experiences of place, their values, perceptions of risk, experiences of the insurers, and decision-making processes. Twenty of these interviews were conducted in 2017 and a further twenty in 2020 with similar questions[1]. Of these forty interviews 35% (n = 14) were renters, 42.5% (n = 17) were males and 57.5% (n = 23) were females. Of those interviewed, 47.5% (n = 19) identified themselves as underinsured or non-insured.

The interviews were conducted either in the homes of the participants or in another place of their choosing, or over video conferencing due to the 2020 Covid-19 pandemic. All the interviews were transcribed verbatim and then analysed using axial coding (Corbin & Strauss 2008). As Holmes (2015, p. 64) suggests, analysing interviews to ascertain how emotional reflexivity has shaped people's thoughts and behaviours is difficult. The entire interview process could be conceived of as a moment of emotional reflexivity where verbal and non-verbal language between the interviewee and interviewer shape how each performs their role in the meeting. Long after the interview is complete, the researcher continues to 'construct respondents' – making choices about which parts of the participant's narrative get highlighted and which are hidden away (Foley 2012, p. 312).

Nevertheless, in this research, careful examination was undertaken of the ways in which participants used language and expressed emotions to portray their social worlds and their relationship with insurance, local geography, and their history (Peck & Mummery 2018, p. 390). While there is no claim of generalisability, it is argued that the research findings presented below contribute to understanding how emotions work in relation to insurance.

Discussion

Emotions and trust in insurers

Public trust in institutions is perhaps at an all-time low and the insurance sector is no different (Tranter & Booth 2019). Indeed, as Giddens (1990) and Beck (1992) argue, uncertainty and expertise create a tension between the public and those professing to provide risk-management, such as insurers. Emotional reflexivity informs and even drives this uneasy relationship. At times, this causes some to be quite angry towards insurers, but they remain insured regardless. It begs the question 'why?,' with Tranter and Booth (2019, p. 205) postulating that it is not simply a matter of trust but of social or institutional norms.

Stories of others are important in the development of emotional reflexivity, as Holmes (2010) describes. This also is apparent as a driver of people's understandings of the insurance sector. Several of the participants in this study talk about 'horror stories' they have heard of people trying to make insurance claims. Milly[2], a middle-aged female homeowner living in the Northern Beaches of Cairns, illustrates this when asked how much trust she invests in insurers;

> That's a tricky one because you *hear all the horror stories*. Suncorp are the cheapest. I have looked around, I have shopped around. They are by far the cheapest but do I read the small print? Would I understand the small print? Probably not [...] I wouldn't say *I totally trust them*. And I think insurance companies are going to try and *get away* with not paying out if they can anyway.

It is important to note here that in the beginning of her response she reflects a certain emotional state in response to stories she has heard. From there, she explains that despite her shopping around for insurers, and her decision to take out insurance, she feels resigned in the belief that they are going to find ways to not pay 'anyway' regardless of her level trust.

People like Milly may be inclined to insure because institutions such as banks require them to do so as part of a mortgage. However, her emotional state is one of resignation. This is exemplified further in the response of a middle-aged female mortgagee named Freya, who lives in the inner-city suburb of Mooroobool. When questioned on her mortgage status she responds, fatalistically, that 'I don't have really any trust [...] if *you've got a compulsory obliged relationship with it*. I do it if I have to.'

Frequently participants call upon 'stories' they have heard that drive their mistrust of insurers. Emotional reflexivity towards insurers is built through experiences others have had, and this extends to thirdhand stories that circulate in the civil sphere. However, broader than this are narratives that emerge through media in the recovery process of large-scale disaster events. In the case of the Brisbane floods of 2011, for example, this included the reporting of stories of individuals finding that they were not covered for flood inundation (von den Honert & McAneney 2011).

Participants in this study at times cited the Brisbane flood event and other disasters such as cyclones Larry and Yasi, as moments where insurers proved that they were untrustworthy. Nell, a middle-aged female homeowner from Edge Hill in Cairns, demonstrates this in the following;

> I've got a *weird feeling* that if it's small scale (the disaster), the insurance companies, if they're going to look at the whole of anything in Far North Queensland, if they've got to fork out a couple of million, yeah they'll take it, they can wear it. But if it is really wide-scale [...] I feel that's when they're going to go, hmm, let's have a closer look and not pay where we don't have to.

Nell's perception of the sector emerges through stories she has heard, and a cultural framing of insurers as distrustful and lacking integrity. Her 'weird feeling' reflects how she interprets past events and possible future outcomes in her suburb. She has an untrusting relationship built through these emotions.

The lack of trust and this narrative that insurers will try and 'get away with it' is reflected in the uneasiness expressed by other participants regarding their insurance policies. Kevin, a middle-aged male homeowner in Edge Hill, illustrates this well;

> There's always that *nagging feeling* you're never going to be covered for the one thing that gets you. And I guess the one thing that bugs me in this area is things like flood and storm surge coverage [...] there seems to be a lot of variation in what's covered and what's not.

Kevin's discomfort causes him to examine and then re-examine his insurance policies to reassure himself that he has the right level of cover. Every renewal period, he questions insurers trying to obtain satisfaction and comfort that risks in his local area are covered. However, despite these efforts, he concedes that risk events like flooding are less 'clear and there seems to be a lot more inter-relationship between the terms and the time triggers as to when they may be covered.' While at the time of interviewing he felt that his chosen insurer was better than prior insurers, he also relays that these nagging emotions of risk-exposure make him feel less 'comfortable' and thus untrusting.

Participants in this study also displayed other emotions including that of anger. One of the major concerns in Northern Australia is the cost of premiums. Insurers claim this is the result of the large reinsurance[3] pool required now in the tropics due to increased risk and danger (ACCC 2019). Despite these claims by insurers, interviewees were not always convinced and expressed anger at what they perceived to be corporate greed. William, a middle-aged male homeowner from the Northern Beaches, gets quite animated when talking about insurance. He argues that insurers 'get away with' what he calls 'price gouging.' In past conversations with his chosen insurer, he relays to me that he was advised that insurance premiums are high due to recent weather events. He responds,

> I'm like horse shit, they can price gouge. The insurance industry I think has taken (a hit) when Brisbane got washed out [...] And they've got to recoup it from somewhere. Now *my feeling is that they don't spread the love*, so they hit the people that it affects more than other people.

William's feelings are not distinct from other participants who accuse the insurance sector of being almost psychopathic in their approach to the insured. One male renter in his mid-twenties from the Northern Beaches, Fredrick (see later), suggests that insurers are 'crooks' and labels them as nothing more than profiteers.

This dim feeling towards insurers is built, as I have argued, through the stories of others that are delivered either in person or through the media. They contribute to participants expressing resignation, dismay, distrust, and even anger. In these cases, emotional reflexivity has already happened. However, in a few interviews, participants exhibited emotional reflexivity as the interview progressed.

Here, the interview process appears to have contributed to them questioning their own insurance policies and associated emotional responses. As an example, Cass, a retired woman and homeowner from Trinity Beach, suggests that before the interview began, she took out her policies and 'started reading it' only to discover 'there's all these exclusions and stuff.' The actual interview itself caused her to wonder 'what will happen if I actually need it (insurance)?' She proceeds when asked 'why are you paying for it?'

> Yes exactly, exactly. I suppose you'd get a little bit but I never expect anything from insurance companies and that's why I'm saying now I'm probably going to cancel my (she pauses)...

It is at this moment that Cass appears guided by her emotions as she reflects on what she was deciding to do. Specifically, she is guided by a fear of victimisation from home break-ins that she has experienced as she suggests that 'I need my home/contents if anything happens because if you get burgled, and that happened to me a few times [...] and I didn't have insurance.' Her emotions of those past events cause her to reflect in the moment and alter her decision. She continues,

INTERVIEWER: How come you don't have confidence in them though?

CASS: Well I heard so many stories about people who now have insurance but then they said oh no but you didn't read the fine print. I suppose I really [...] should go into the details [...] (but) you do it the first time and then, yeah you just keep repaying.

While participants above relay some element of emotional reflexivity in their understanding of the sector prior to the interview, this dialogue with Cass demonstrates how emotions in the moment inform her decisions to stay insured. Like Booth and Harwood's (2016, p. 49) 'insurantial moment' where 'in different contexts, different rationales are applied' to deal with specific 'uncertainties,' Cass moves from considering withdrawing from insurance, to confirming her desire to hold a policy based on different risk contexts.

Emotions and risk-taking in non-insurance

Insurance is emotional. However, thus far the chapter has focused on areas where those emotions are filled with negative feelings, especially towards the insurers themselves. In this section, I want to relate some of the feelings of complacency and optimism but also excitement that emerge in discussions around risks and how this relates to insurance. When it comes to risk assessment, not all share in the type of dread that Beck (1992) suggests. There are those, as identified by Lupton (2013), that draw on risks to heighten their emotional states. In reflecting on the risks of living in the Cairns region, for instance, one mid-twenties male renter, Fredrick, relates that although cyclones are a major threat, he feels they are also just a bit of 'fun.' When pressed why he feels that way, he adds, 'you go to your friends who lives in a good house and get on the piss (drink alcohol).' Fredrick is referring here to gatherings called 'cyclone parties' where people come together, bunker down in a solid structure and drink together. He relates further,

It's just there's a buzz about the place (when a cyclone approaches). When it's happening. I've always been told oh, *it's been alright*, you go to a small room of the house and you sit it out, take the windows off. There's a social solidarity during, before and after a cyclone *which is exciting*. As you know we don't have too much of that quite often.

Fredrick's feelings are punctuated by emotions of excitement. These are derived from his own, and others, past experiences where the thrill of taking time off normal duties and sitting together creates anticipation of fun and enjoyment for him. Even though Cairns has not experienced a severe cyclone event for several years now, he feels confident but also looks forward to these risks with excitement.

These emotions cause Fredrick to reject insurance for his possessions as a necessary part of his life. He does not insure his contents at all and suggests he feels 'quite pessimistic' regarding how insurance plays such a 'primary role in society.' It might be added that Fredrick also 'just lives with' risks arguing that, 'it's a good way to live.' While Fredrick uses these emotional states to negate any desire to insure his contents, other participants use confidence and a sense of fatalism to do a similar thing. Gemma, a middle-aged single mother of three from the Northern Beaches area, exemplifies an attitude that was prevalent amongst renters. When discussing the need for insurance, she dismisses it with a wave of the hand arguing that 'we live a very simple, non-materialistic life, so for us, I don't feel that we need the insurance to cover.' She adds later,

INTERVIEWER: how confident would you be that you would be ok after (a large-scale natural disaster)?
GEMMA: I've done it – when I left home, we sold everything we owned and we put our possessions in the car and we drove here and we started a new life. So for me, we would do it again. It would be difficult but *we don't live that materialistic life* for that to be a big, big problem.

Despite having school-aged children and multiple items in her home, such as a TV, beds, clothing, refrigerator, freezer, Gemma expresses confidence and exhibits feelings of optimism in her ability to start again without insurance. Her experiences in the past allow for an emotional response best described as complacence. Later in the interview when pressed on the risks of cyclones, she responds with a sense of fatalism, 'if a cyclone comes, do what I can do. Whatever happens, happens […] my belief is, when it's your time, it's your time.'

This 'fatalism' is a recurring theme amongst some of our participants, even those who are insured. These are also the kinds of insights that inform critiques of Beck's (1992) work wherein 'public perceptions of risk' are not responded to always through 'rational choice' or cost-benefit calculations (Mythen 2005, p. 143). Rather, risks can at times be influenced by 'fate in patching together personal interpretations of danger' (Mythen 2005, p. 143). As described above in the case of Gemma, there is a psychological moment, built on past experiences, that allows her to wave off potential dangers and allowing her to have confidence in herself despite unknowns and uncertainty (Osbaldiston, McShane and Oleszek 2019). Frequently, it is the renters in this study who are also dismissive of insurance because they feel little attachment to their material possessions, claiming them to be of little to no value. As one middle-aged male renter from within Cairns' internal suburbia argues, 'we don't really have much value to begin with.' Thus, distinct from other participants, the decision not to insure is based on low emotional connection to an imagined future sense of loss. Potential peace of mind provided by insurance is dismissed and negated.

Conclusion

As Holmes (2010, p. 141) argues, our modern lives are increasingly punctuated by 'uncertainty' making 'rational choices based on the probability of certain outcomes unfeasible.' The modern age requires individuals to become reflexive, consistently judging between choices and weighing up decisions according to the multiple information sources made available to them (Giddens 1990). The difficulty in this framing is that the 'reflexive self is formed by emotional relations to others and thus emotions play a more complex part in deliberations' (Holmes 2010, p. 142). In this chapter, I have attempted to show how emotions play a role in the positioning of the reflexive self with insurance in the context of cyclone risks.

In particular, the ongoing configuration of the individual in relation to their decision to insure is punctuated by emotions. These feelings are fed by distrust of insurers along with a desire to insure in the first place. As shown above, those who feel a social and moral obligation to insure have little choice but to do so (Tranter & Booth 2019). This relationship is one best understood regarding emotions, not simply rational, cognitive framings. At times this is deep distrust that the insurer will meet their obligations. At other times, the desire for peace of mind counters any decision to cancel insurance or not insure. Furthermore, feelings of optimism and confidence based on past experiences with natural disasters or major life-disrupting events, especially for renters, can convince individuals that they do not need insurance.

To conclude it is worthwhile noting that Holmes (2015, p. 64) herself has argued that studying emotional reflexivity empirically can be quite difficult. As shown above, there are traces of emotionality invested into the narratives of these participants, but these are perhaps tenuous. The interview with Cass (above), however, exemplifies how emotions play a role in a decision to insure or not. In future research, it appears worthwhile drawing these out further by exploring specific events where participants are invited to reflect on different scenarios, as Cass did in her own mind. In doing so, we might be able to tap into the emotions that guide and even underline different reflexive positions that people adopt into their everyday lives.

Notes

1 The 2017 interviews formed part of an internally funded project to James Cook University entitled 'Understanding under/non-insurance in the Cairns Northern Beaches.' The 2020 interviews formed part of a consultancy undertaken for the Cairns Regional Council entitled 'Building community economic capacity through understanding insurance levels.'
2 All respondents have been given pseudonyms to protect their identities.
3 Reinsurance is the practice of insuring insurers. Reinsurers provide cover to protect insurance companies from becoming insolvent due to a large-scale event that drains the resources of the insurer.

References

Australian Competition and Consumer Commission (ACCC) 2019, *Northern Australia insurance inquiry, second interim report*, ACCC, Canberra.

Beck, U 1992, *Risk society: Towards a new modernity*, Sage, London.

Booth, K & Harwood, A 2016, 'Insurance as catastrophe: A geography of house and contents insurance in bushfire-prone places', *Geoforum*, vol. 69, pp. 44–52.

Booth, K & Kendal, D 2020, 'Underinsurance as adaptation: Household agency in places of marketisation and financialisation', *Economy and Space*, vol. 52, no. 4, pp. 728–746.

Burkitt, I 2012, 'Emotional reflexivity: Feeling emotion and imagination in reflexive dialogues', *Sociology*, vol. 46, no. 3, pp. 458–472.

Corbin, J & Strauss, A 2008, *Basics of qualitative research: Techniques and procedures for developing grounded theory*, Sage, London.

Foley, L.J 2012, 'Constructing the respondent', In Gubrium, J.F, Holstein, J.A, Marvasti, A.B & McKinney, K.D (eds.), *The Sage handbook of interview research*, pp. 305–316, Sage, London.

Giddens, A 1990, *The consequences of modernity*, Stanford University Press, Palo Alto, CA.

Hogarth, R.M, Portell, M, Cuxart, A & Kolev, G.I 2011, 'Emotion and reason in everyday risk perception', *Journal of Behavioural Decision Making*, vol. 24, pp. 202–222.

Holmes, M 2010, 'The emotionalization of reflexivity', *Sociology*, vol. 44, no. 1, pp. 139–145.

Holmes, M 2015, 'Researching emotional reflexivity', *Emotion Review*, vol. 7, no. 1, pp. 61–66.

Lowenstein, G.F, Hsee, C.K, Weber, E.U & Welch, N 2001, 'Risk as feelings', *Psychological Bulletin*, vol. 127, no. 2, pp. 267–286.

Lupton, D 2013, *Risk*, 2nd Edition, Routledge, New York.

Mythen, G 2005, *Ulrich Beck: A critical introduction to the risk society*, Pluto Press, London.

Osbaldiston, N, McShane, C & Oleszek, R 2019, 'Underinsurance in cyclone and flood environments: A case study in Cairns, Queensland', *Australian Journal of Emergency Management*, vol. 34, no. 1, pp. 41–47.

Peck, B & Mummery, J 2018, 'Hermeneutic constructivism: An ontology for qualitative research', *Qualitative Health Research*, vol. 28, no. 3, pp. 389–407.

Rick, S & Lowenstein, G 2008, 'The role of emotion in economic behaviour', In Lewis, M, Haviland-Jones, J & Barrett, L (eds), *Handbook of emotions*, pp. 138–156, The Guilford Press, London.

Sjoberg, L 2007, 'Emotions and risk perception', *Risk Management*, vol. 9, pp. 223–237.

Slovic, P, Peters, E, Finucane, M.L & MacGregor, D.G 2005, 'Affect, risk and decision making', *Health Psychology*, vol. 24, no. 4, pp. s35–s40.

Tranter, B & Booth, K 2019, 'Geographies of trust: Socio-spatial variegations of trust in insurance', *Geoforum*, vol. 107, pp. 199–206.

von den Honert, R & McAneney, J 2011, 'The 2011 Brisbane floods: Causes, impacts and implications', *Water*, vol. 3, pp. 1149–1173.

13 Insure the volume?

Sensing air, atmospheres, and radiation in the Chornobyl Exclusion Zone

Christine Eriksen and Jonathon Turnbull

Introduction

This chapter examines a novel and intriguing occurrence concerning recent wildfires in the Chornobyl[1] Exclusion Zone (CEZ) in Ukraine, which rerelease radioisotopes originally deposited into local ecosystems during the Chornobyl nuclear accident of 1986. In particular, we focus on the wildfires that burned in the CEZ and elsewhere in Ukraine in April 2020, blanketing the capital, Kyiv, in smoke. The chapter considers what this kind of uncontainable, airborne, and hazardous phenomenon potentially means for insurance, as well as insurance and disaster research.

Examining this phenomenon in the context of insurance is somewhat paradoxical. Traditionally, to make an event insurable there has to be a clear causal event. Yet, how do you calculate insurance, let alone determine liability, for a harmful event (the Chornobyl disaster) that occurred decades ago by an actor (the Soviet Union) that no longer exists and can no longer be held accountable? As we demonstrate in this chapter, the original volume of damage in 1986 has expanded over time with the leakage and seepage of radioisotopes into the ground, out into waterways, and up into the air via natural processes. This creates a problematic dilemma for governance and insurance. It points to a spatiotemporal problem inherent to insurance studies. We attempt to address this problem with the help of a theoretical framework that draws on recent scholarly work in the social sciences, particularly in human geography, to foreground affective atmospheres, sensing assemblages, and volumetric sovereignty. Rather than pointing specifically to what might be insured in this context, we aim to problematise the very notion of insuring against disasters in the techno-natures of the Anthropocene.

In this chapter, we argue that affective atmospheres – not insurance frameworks – governed how people reacted to and managed the wildfires and smoke in 2020. The tangible and intangible risks of the rerelease of radioisotopes by wildfire in the CEZ are an example of the long-term consequences of 'high-tech risks':

DOI: 10.4324/9781003157571-17

In the afflictions [high-tech risks] produce they are no longer tied to their place of origin – the industrial plant. By their nature they endanger *all* forms of life on this planet. The normative bases of their calculation – the concept of accident and insurance, medical precautions, and so on – do not fit the basic dimensions of these modern threats. Atomic plants are accident no more (in the limited sense of the word 'accident'). They outlast generations... This means that the calculations of risk as it has been established so far by science and legal institutions collapses. Dealing with these consequences of modern productive and destructive forces in the normal terms of risk is a false but nevertheless very effective way of legitimizing them.

(Beck 1992, p. 22)

The insurance industry plays an intrinsic role in the modern-day risk society described in Beck's seminal work, which was originally published the same year as the Chornobyl disaster. Here, 'insurance operates a security technology that, while ascribing value to life, capitalises livelihoods, and promotes [certain] lifestyles' (Lobo-Guerrero 2011, p. xi). Yet, due to the difficulty of calculating the likelihood and magnitude of environmental hazards, especially those with effects at a global scale, certain events escape the logic of insurance and are rendered 'catastrophic' anomalies. Since the 1990s, innovative insurance technologies have changed this through two manoeuvres: parametric insurance and the securitisation of catastrophic risks in global markets by means of financial derivatives (Lobo-Guerrero 2011; Grove 2012; Collier 2013). Insurance products, such as 'catastrophe bonds' still rely on the design of predictable variables that are difficult, if not impossible, to meet in the context of radioisotopes rereleased by wildfires: 'observable and easily measurable, objective, transparent, independently verifiable, reportable in a timely manner and stable and sustainable over time' (Lobo-Guerrero 2011, pp. 83–84). Politically and economically, no one wants to insure the leaky and ungovernable radioactive particles in the CEZ.

As Lobo-Guerrero (2011, p. 78) highlights, the 'crossbreeding of insurance and capital markets ... [is] related to a wider process of governing through assemblages of risk.' This resonates with Beck's (1992, p. 21) understanding of risk as the 'systematic way of dealing with hazards and insecurities induced and introduced by modernisation itself.' Our case study, however, highlights the importance of a different set of assemblages in managing risk, namely sensing assemblages. The diverse and sometime overlapping forms of sensing used for detecting radiation, such as the technologies used in Kyiv and the bodies of firefighters, highlight the importance of thinking with volume to understand affective atmospheres. These affective atmospheres, together with the Anthropocene's proliferating techno-natural ecosystems, fundamentally challenge some of the legal terms traditionally used to define liability, such as labelling a natural hazard outside of human control, for which no person can be held responsible, an 'Act of God.'

The historical premise of our analysis is the tangible lack of insurable relationships in the Soviet Union at the time of the disaster in 1986. Today, operators of nuclear power plants are liable for any damage caused by them, regardless of fault. This was not the case in 1986. The Soviet Union, moreover, was not a signatory of the conventions that addressed international liability issues: The Paris Convention on Third Party Liability in the Field of Nuclear Energy of 1960, the Vienna Convention on Civil Liability for Nuclear Damage of 1963, and the Brussels Supplementary Convention of 1963 (IAEA 2020). Thus, aside from the estimated 300,000 evacuees that had lived within what became the parameters of the CEZ (who were relocated and partly compensated by the state), compensation was difficult to claim for people affected by the disaster in the acute aftermath and the ensuing years. People were required to prove the link between radiation-related exposure and the health impairments they were experiencing. This continues to pose a major challenge for many people affected by the Chornobyl disaster. The effects of radiation are often subtle, manifesting stochastically, long after the exposure event. The intergenerational effects of exposure are also a contentious issue. The lack of quality health care in the economically and politically unstable post-Soviet states compounded these issues by making it difficult, especially for poor people, to receive compensation. Petryna (2016) outlines how this led to a new form of 'biological citizenship' in which Chornobyl suffers mobilised assaults on their health to stake claims for biomedical resources, social equity, and human rights. As Davies (2015, p. 230) writes, people felt abandoned or exposed twice, 'once to the hidden threat of radiation, and once more to a state that has abandoned them.'

There have been significant improvements to national laws and international conventions since then, first in response to the Chornobyl disaster, and subsequently after the Fukushima disaster in 2011 (World Nuclear Association 2018). However, in 1986, people in the Soviet Union were at the mercy of political will for any form of compensation, while the consequences felt overseas fell back on national principles of common or civil law as well as political will (Schwartz 2006). The strict liability of the nuclear operator, which is one of the key principles of most conventions and laws regarding nuclear third-party liability today, is significant because an insurance claimant does not have to prove how an accident occurred (i.e. prove fault) (World Nuclear Association 2018). However, even with the extension of the prescription/extinction period of insurable damage to 30 years post-event in the revised international conventions, many impairments to the environment or personal health are not covered retroactively because of the difficulty of proving causality years or decades after a disaster (Schwartz 2006). Complicating matters further in the context of wildfires in the CEZ, is the lack of an international legal agreement on the vertical extent of sovereign airspace (Billé 2017).

In what follows, we begin with a description of our theoretical framework before introducing our case study. We then build on the growing body of scholarly work on affective atmospheres, sensing assemblages, and volumetric sovereignty, which underpin our theoretical framing, to examine the ungovernable atmosphere and the uncontrollable flow of air and radioactive particles in the CEZ. We conclude by suggesting the implications of such thinking for critical disaster and insurance studies more broadly.

Atmospheric frameworks

Atmosphere as an affective phenomenon has become a critical concept in the last decade, allowing scholars in the social sciences 'to grasp the affective materiality of spacetimes that are diffuse and excessive of bodies yet also palpable through the sensory capacities of those bodies' (Engelmann & McCormack 2018, p. 187). Affect concerns the precognitive, felt, impulsive, ebbs and flows of lived experience. On these terms, atmosphere does not only refer to the envelope of gases that surrounds the Earth, but also to a field of affective experience. This mode of thinking helps us make sense of the 2020 wildfires in the CEZ, their representation, and sensed embodiment. Following Choy (2018, p. 56), it is not 'the actuality of atmospheric objects or worlds, but rather how people adjust relations and relational capacities when motivated by an atmospheric question,' which determines how affective atmospheres are sensed, embodied, and only later expressed in emotional terms (see also, Anderson 2009; Anderson & Ash 2015). The qualities of physical atmospheres, such as temperature, pressure, and humidity, vary over space and time. Human bodies can often sense these variations. Yet, this is not always the case. For example, the invisibility of radiation, and the human body's inability to detect low doses, means atmospheric radiation (and other atmospheric pollutants) gain some of their affective capacities through participating in sensing assemblages. Here, 'non-human bodies and devices of various kinds,' with the capacity to be 'affected or perturbed,' come together to render them present to human senses (Engelmann & McCormack 2018, p. 188). Recent work on environmental sensing has probed the expansive scope of sensing technologies that challenge human conceptions of agency, experience, and the boundaries of the body (e.g. Gabrys 2020). These forms of sensing, observing, and marking pertain to the wildfires in the CEZ.

Like with atmosphere, a growing body of scholarly work concerning volumetric and voluminous conceptualisations of space and sovereignty is helpful for our analysis of the spatiotemporal problem inherent to insurance studies. Largely inspired by Sloterdijk's (2009) notion of 'spheres' and Weizman's (2002) 'politics of verticality,' these recently developed approaches enable three-dimensional (3D) understandings of space, which challenge the often disembodied two-dimensional (2D) understandings of elemental relationships (Adey 2015; Billé 2017; Billé 2019). Elden's (2009; 2013; 2017) work has been particularly influential, initially emphasising the economic, strategic, legal, and technical dimensions of vertical politics, and later acknowledging the equal importance of materiality and embodiment in thinking with volumetric spaces (Elden 2020). The title of this chapter is a play on Elden's (2013) term 'secure the volume.'

Thinking space volumetrically involves thinking through 'height and depth instead of surface, [and thinking in] three dimensions instead of areas' (Elden 2013, p. 35). Volumetric approaches have largely been concerned with metrics and measurement, whilst voluminous approaches emphasise embodied understandings of material entities (e.g. dust) within dynamic assemblages (e.g. air, oceans), which are never static and cannot be contained (Steinberg & Peters 2015). This distinction is important when considering the issues of governance and liability intrinsic to volumetric sovereignty, as it helps us unravel why a 2D conceptualisation of space is inadequate for mapping the leaky materialities of the CEZ. As Steinberg and Peters' (2015, pp. 258–259) highlight,

> ...legal institutions will always attempt to delimit volumes into strata just as they will always attempt to delimit horizontal spaces into areas [...] But the nature of territory as a political technology means this process will always be met with a resistance that reflects underlying dynamics that are both social and geophysical. [... In turning] our attention to the volumes within which politics is practised and territory is produced we must continually rethink the borders that we apply to various materialities and their physical states.

In bringing together these bodies of work, we aim to show how volumetric space and the movement of (contaminated) air are bound up with ideas and information as well as sensing and feeling (Mitchell 2011). These concepts are pertinent to insurance studies, as they speak to the difficulties posed by complex configurations of causation, harm, and responsibility that characterise disasters in the techno-natures of the Anthropocene. They attune scholars to the muddled spatialities and temporalities that come with accelerating environmental change. In what follows, we foreground the relationships between affective atmospheres, sensing assemblages, and volumetric sovereignty to illustrate why the volume of radioactive particles rereleased by wildfires in the CEZ is difficult to manage and even harder to insure.

Wildfires in the CEZ

On 26 April 1986, the explosion and subsequent open-air graphite fire at Reactor No. 4 of the Chornobyl Nuclear Power Plant (ChNPP) contaminated the soil, water, and atmosphere alike with radioactive material at a rate equivalent to twenty times that released during the 1945 atomic bombing of Hiroshima (Higginbotham 2019). For nine days, the fire produced updrafts that lifted deadly plumes of radioactive material into the atmosphere, which wind and rain distributed over (what was then) the western parts of the Soviet Union, particularly Belarus, Ukraine, and Russia, as well as many European countries.

The area most heavily contaminated is known as the Red Forest – a 10-km^2 area surrounding the ChNPP, named after the pine trees that turned a ginger-brown colour after absorbing high levels of radiation. The ensuing clean-up operations bulldozed the majority of the pine trees and buried them, together with other contaminated materials, in newly dug trenches (WHO 2006). These unlined and leaky trenches were then covered with a thick carpet of sand, and pine plantations were replanted in an attempt to soak up radiation and prevent its spread into the groundwater. Up to 85 percent of the radioactivity in the Red Forest is concentrated in the soil, with the remainder deposited in the bark, needles, timber, and branches of the remaining trees and other forms of vegetation (Hao et al. 2009). The gradual and indefinite evacuation of over 300,000 people and the abandonment of agriculture within the CEZ has, with time, facilitated the presence of a diverse range of flora and fauna, despite the radioactive fallout they absorb, eat, or inhale (Mycio 2004). Nevertheless, the area remains one of the most contaminated regions in the world today (Brown 2019).

The relatively undisturbed growth of vegetation since 1986 has resulted in another problem. Over a thousand wildfires have burnt inside the CEZ since it was established, and in April 2020, fires in the area surrounding the ChNPP became a worrisome presence once again. Whilst the spectres of 1986 are alive in these wildfires, they pose different threats to the original disaster in scale and intensity. The graphite fire of 1986 released a huge amount of radioactive material high into the atmosphere that was distributed globally. The 2020 wildfires released clouds of smoke containing radioactive particles and mineral dust from radioactive pollutants absorbed and held over time by vegetation and soil. The smoke enveloped surrounding areas, including Kyiv, 100 km to the south, where one of us (Turnbull) happened to live, directly experiencing (inhaling) the smoke.

These wildfires and the drifting smoke are cause for international concern, as their likelihood increases with climate change (Amiro et al. 1996; Eriksen 2022). Whilst small increases in radioactivity were detected in the air in Kyiv in 2020, an air filter station in the north of Norway registered an increase in the presence of Caesium-137 (one of the most common radioisotopes at Chornobyl), which potentially stemmed from the CEZ (Nilsen 2020). Scientific assessments highlight that inadequate forest management 'has resulted in a high wildfire hazard in the 260,000 ha of forests and grasslands of the Ukrainian part' of the CEZ (Zibtsev et al. 2015, p. 40). Given the right wind conditions, smoke and dust can travel across not just geopolitical borders but across continents due to the indiscriminate crossing of borders by atmospheric particulates moved by uncontrollable forces, such as wind and air currents (Eriksen and Ballard 2020). This poses a health threat both to the people who eat food grown in fallout areas, and to people who inhale contaminated smoke, like firefighters and land-stewards who attempt to manage the wildfire threat. A radiological assessment of two sizeable wildfires in the CEZ in April and August 2015 found that the (effective)[2] radiation doses were above 1 millisievert (mSv) per year within the borders of the CEZ, but outside and in the rest of Europe doses were much lower, 'equivalent to a medical X-ray image at most' (Evangeliou et al. 2016, n.p.). For context, 100 mSv per year is the lowest level of exposure at which there is clear evidence of an increased risk of cancer (IAEA 2014).

In the next section, we examine the lingering consequences of the Chornobyl disaster in the context of the atmospheric challenges posed by the rerelease of radioactive fallout by wildfires that, according to local reports, were ignited by neglectful farmers and/or an arsonist. We interrogate how the relationship between atmospheric sensing, radioactive air pollution, wildfires, and volumetric sovereignty renders this ongoing socio-technical disaster uninsurable.

Sensing air, atmospheres, and radiation

Wildfires are a yearly, common occurrence in Ukraine, mostly because of farmers (sometimes illegally) burning fields, which get out of control. For this reason, the smoke that enveloped Kyiv for over a week in April 2020 was not solely from wildfires in the CEZ. The indeterminable origin of the smoke gave it an eerie quality: Was it or was it not radioactive? This concern preoccupied state agencies and others who decided to check radiation levels via spectacular acts of atmospheric sensing. These acts involved taking atmospheric measurements in public, including on Kyiv's main street Kreshchatyk, which runs through Maidan Nezalezhnosti (Independence Square). Kreshchatyk was the location of the May Day Parade in 1986 where Soviet authorities, five days after the Chornobyl disaster, knowingly exposed members of the public to high levels of radiation as the radioactive cloud enveloped them with particles carried on air currents from the ChNPP. Attempting to sense radiation today in this symbolic place is a spectacular and haunting reminder that the Chornobyl disaster continues.

We spoke with a radiation expert, Boris[3], in December 2020 to determine the purpose of the testing on Kreshchatyk earlier in the year. He suggested that the tests were a publicity stunt and that the equipment used was inappropriate for measuring the kind of exposure that might actually be a threat in this area. Boris told us that microscopic pieces of radioactive material known as 'hot particles' were a cause for concern, and although rare, may be very dangerous when ingested. The effects of hot particles are controversial in the scientific community. If ingested, some scientists suggest, these radioactive materials can settle in specific tissue in the body and deliver a concentrated dose of radiation to a small group of cells. He also suggested that the 'experiments' were carried out in places where radioisotopes were unlikely to exist *if* they had made the atmospheric journey to Kyiv. As with the initial silence and then denial by the Soviet authorities in 1986, these spectacular acts of sensing in 2020 were not useful in determining the unequal distribution and effects of the potentially radioactive air pollution. This is despite the fact that air pollution 'presses the question of how atmospheric things disperse and accumulate in unequal concentrations' (Choy 2020, p. 105).

For Boris, the atmospheric sensing in Kyiv was not strictly a scientific act, but rather a political one. As noted by Choy (2018, p. 57) 'numbers and measures themselves are both visceral and affecting.' The chances of registering an increased dose rate outside the CEZ was unlikely, as the 2020 wildfires were not energetic enough to pick up and carry radioisotopes long distances. This is unlike the original disaster during which vast releases of energy allowed radioisotopes to travel across continents (Higginbotham 2019). This spectacular sensing, therefore, was generative of the affective atmosphere associated with the wildfire, as the readings taken on Kreshchatyk were posted on social media and circulated widely.

During the wildfires, residents of Kyiv had to close their windows to keep out the smoke as it engulfed streets and apartments. Living in Kyiv meant embodying, and sharing with neighbours, 'the specificity of experiences of being in and indeed of witnessing things becoming airborne' (McCormack 2009, p. 27). As Kyiv residents stayed indoors, a sense of 'being contained' arose, which was especially troubling for people whose movements were already restricted due to the COVID-19 lockdown. The lack of fresh air both inside and outside gave rise to a sense of envelopment by air pollution. Whilst unpleasant at the best of times, the fact that the smoke was potentially coming from the CEZ added another layer of fear. Such embodied everyday responses to the affective atmosphere of the CEZ wildfires are important to understand in the context of governing nuclear spaces. Nuclear wildfires reanimate a particular imagination associated with nuclear fallout from earlier disasters. Be it fallout from the Chornobyl disaster or nuclear weapons testing elsewhere (Masco 2006; Eriksen 2022), the affective atmosphere created by wildfires involves uncertainty and fear. Because of radiation's ambiguous status in public discourse, these atmospheres form regardless of the actual threat posed.

Much of the maintenance work to contain both the fear and threat is carried out by firefighters who, while tackling wildfires in the CEZ, simultaneously mollify public concerns about exposure. Yet, each time they do so, they are individually exposed to contaminants that persist in the landscape long-term as a result of the original disaster. Firefighters can reach the limit of what is considered a safe annual radiation dose over a 'relatively small number of days' (Zibtsev et al. 2015). While radiation levels during a wildfire are not as high as they are during a direct encounter with a radioactive source, the cumulative effect of exposure to low levels of radiation over the career of a firefighter is unknown. Dose rates are not always the most useful measure for understanding the possible consequences of such atmospheric exposures according to some. In areas where radiation spikes are recorded in proximity to the wildfires, 'hot particles' (as discussed above) are made airborne and can be inhaled. There is concern that radiation risk models derived only from external exposure, do not account for the risks associated with internally ingested hot particles. Background (or ambient) radiation levels only represent an average exposure. As such, '[t]he true evaluation of nuclear risk is tied to specific exposures rather than the background radiation count' (Masco 2006, p. 299). The effects of hot particles are difficult to measure, though, as they operate at extremely small scales. They complicate the notion of how we experience the environment in such miniscule quantities (cf. Creager 2018), pointing to how the environment is not just outside us, but also within us (Alaimo 2010).

Furthermore, radiation readings and sensors alone cannot account for the effects of exposure to radiation during wildfires. Rather, sensing radioactivity is an embodied experience for firefighters who lack adequate monitoring and safety equipment and instead feel a tingling sensation on their skin when fighting a fire in a radioactively contaminated area (Evans 2011; Eriksen & Ballard 2020; Eriksen 2022). As Shapiro (2015, p. 375) notes, '[b]odies are sensors that indicate the presence of toxicants and, in some cases, specify their atmospheric concentration with uncanny precision.' Creager (2018, p. 70) similarly suggests, 'human bodies come to serve as unconscious sensors of their environments.' A range of animals are also enrolled as sentinels for determining the effects of the Chornobyl disaster on 'nature' more broadly (Petryna 2013; Masco 2006). These vernacular accounts of radioactive landscapes contribute to the sensing of 'nuclear weather-worlds' (Alexis-Martin 2020). Human and nonhuman bodies, through microscopic atmospheric encounters, 'are often embroiled in sensing the world well before cognition catches wind of protracted chemical encounters' (Shapiro 2015, p. 375).

Atmospheres, then, are something we are immersed in. They are embodied, but they are also sensed and represented. These representations contribute to the overall affective atmosphere of atmospheric things, such as potentially radioactive smoke. Thinking atmospherically draws attention to the ways in which wildfires are atmospheric from the outset. Wind and oxygen fanned the spread of the 2020 wildfires in the CEZ, while muchneeded rain ultimately extinguished them. In attending to the affective and material atmospheres of these wildfires, we come to understand their representation and sensed embodiment as issues of suspension: material and affective. First, wildfires resuspended the radioisotopes originally deposited during the disaster in 1986, causing problems for the firefighters exposed to them. Second, the clouds of smoke that spread from the CEZ and enveloped surrounding areas may or may not have been radioactive, suspending exposed populations in a liminal state of uncertainty. This state of uncertainty is compounded by the stochastic nature of radioactivity's effects – especially hot particles – which are difficult to account for and measure. The atmospheric effects of the wildfires, therefore, are both material and affective.

What, then, do notions of suspension, which challenge understandings of the terrestrial and the atmospheric as separate, do to notions of sovereign power and accountability for environmental catastrophes? What are the implications for thinking about insurance in the context of disasters that unfold in uncontainable spaces?

Volumetric sovereignty and leaky materialities

Measurement and containment – or rather the lack thereof – is key to understanding the long-term and uninsurable consequences of the Chornobyl disaster. Notions of the CEZ as a contained space are complicated when we think volumetrically and voluminously. Understanding the CEZ as a vertical space that extends both upwards and downwards allows us to reframe the movement of radioisotopes beyond standard 2D cartographic representations of contamination. How radioisotopes evade borders is increasingly understood through their embodied existence in local ecologies, biologies, and ecosystems (Brown 2015; Brown 2019). Moreover, radioisotopes from Chornobyl continue to leak into consumable products that are consumed locally *and* globally (Brown 2020; Davies 2015). Radioisotopes deposited in the soil are absorbed by vegetation, becoming embodied in bark, leaves, grass, fruit, and berries, amongst other things. In turn, animals, birds, and insects ingest them, contributing to the ecological movement of radioisotopes through their own vertical and horizontal mobilities. When wildfires occur, the radioisotopes are released via combustion into the atmosphere where they circulate and, depending on the atmospheric and particulate conditions, can drift with wind and rain across geopolitical boundaries. This poses problems for determining accountability in relation to industrial accidents that are not contained spatially and temporally.

Radioisotopes also leak and seep from the soil and unlined trenches into the groundwater and waterways. As with creatures, they become embodied in people when the polluted water, vegetation, and air are consumed or inhaled. Adequately mapping the leakage that has occurred in the decades since the Chornobyl disaster is impossible, as radioisotopes are metabolised through complex socioecological systems. As Cons (2017) highlights, 'seepage is a process, not an event. It is the ooze that heralds the failures of projects aiming to produce space and territory as solid containers.' In attempting to 'secure the volume,' Elden (2013) states the need to know where the law applies (and ceases to apply), and which law is operable. Yet, this approach leaves any effort to secure or insure the volume of radioactive smoke particles released by CEZ wildfires in limbo, as there is no international legal agreement on the vertical extent of sovereign airspace (Billé 2017). This creates a problematic horizontal dilemma for governing volumetric sovereignty.

The inability to control the movement of particulates – be they laced with radiation or not – is a poignant example of the inadequacy of existing legal frameworks to address the leaky materialities of the CEZ. Reflecting on Zee's (2017) study of dust storms travelling from China to South Korea, Billé (2017) suggests, '[d]ust blurs the line between earth and liquid as it is driven downwind. As a fugitive substance, it is a voluminous entity that surrounds, embraces, confuses, and potentially kills.' In the case of the CEZ, dust particles are one of the main parameters that underpins the 25-year multinational effort that went into funding and constructing the gigantic shelter that now covers the leaking and crumbling original sarcophagus over Reactor No. 4. This effort is already (possibly inadvertently) thinking with volume as they grapple with the consequences of a potential rerelease of radioactive materials. These voluminous, leaky materialities are an issue of international concern, particularly for more affluent European countries downwind of Chornobyl who have the means to mitigate the threat. Yet, it is also the crux of the uncertainty embodied by firefighters in the CEZ; an issue, as argued in the previous section, laden with affective atmospheres as well as leaky materialities.

Thinking with volume, in conversation with human geography approaches to affective atmospheres, complicates ideas of territory, sovereignty, and property, making it difficult to insure against certain events, such as radioactive air pollution resulting from the rerelease of nuclear fallout (see also Goldstein (2020) on the Southeast Asian haze crisis). If lightning or spontaneous combustion had lit the wildfires in the CEZ (i.e. natural causes), they could be deemed an 'Act of God.' Yet does this actually apply in the techno-natural ecosystem of the CEZ? The wildfires exacerbate an existing socio-technical disaster, rereleasing radioisotopes that pollute the surrounding air. Accountability for the radioactive air pollution is complicated further, as the wildfires were a result of negligent farming practices or deliberate arson. While there may be an individual (or group of individuals) at fault for igniting a fire that got out of control, they are not to blame for the long-term consequences of the irresponsible actions of the Soviet authorities 36 years ago. Holding someone accountable (even if only by issuing a fine), or labelling a wildfire as a purely natural event, enables 'those in power' to cover up, or even justify, decisions and actions that have 'proved both environmentally unsound, and socially, if not morally, bankrupt' (Steinberg 2006, p. xiv). As with many other events particular to the Anthropocene, wildfires in the CEZ fundamentally challenge what counts as an 'Act of God.'

Conclusion

Air currents, like water currents, are carriers of change. Yet, ownership and responsibility are opaque when it comes to what these currents carry and the consequences of the change they deliver. Once radioisotopes are rereleased by wildfire, moving downwards into the soil and upwards into the atmosphere, the dimensionality implied by 'volume' and the calculability implied by 'metric' (Elden 2013) easily becomes abstract and dematerialised (a point also emphasised by Steinberg and Peters (2015) in relation to oceanic materiality). Such abstract and dematerialised forms of reimagined and reanimated disasters, which originally occurred decades ago, are incompatible with standard geographical approaches to territorial boundaries and the legalities of insurable matter. Yet, the consequences of these voluminous, leaky materialities are unequivocal (a point also demonstrated by Nading (2017) in the context of chemicals). In the CEZ, where first the graphite fire and then wildfires burned, and in the surrounding areas where radioactive smoke continues to drift, radiation has continuously reshaped local ecologies, biologies, and ecosystems in acute and subtle ways since 1986.

Thinking with atmospheres, sensing assemblages, and volume, we suggest, offers fruitful avenues for critical disaster and insurance studies to engage with events that refuse containment, and that invoke affective responses at scales unbound by geopolitical territories, sovereign states, and insurance technologies. In this chapter, we introduced such literatures to both enliven and materialise scholarly discussions of insurance, and to understand the long-term lived experiences and continuing consequences of the Chornobyl disaster. Engaging with work on atmospheres as both material and affective, we outlined how atmospheric sensing became a politicised and spectacular event in the wake of the 2020 wildfires in the CEZ. Attending to sensing assemblages, we showed how the bodies of firefighters are emblematic of the embodied and uncontainable effects of radioactive pollution. Turning to work on volumetric sovereignty, we highlighted how 2D conceptualisation of space are inadequate for mapping the leaky materiality of the CEZ, especially as the 2020 wildfires re-suspended radioisotopes atmospherically. Conceiving of space in 3D allowed us to challenge notions of insurability and sovereign space. As Elden (2009, xxii) suggests, '[r]ecognizing the vertical dimension of territory shows that territory is a volume rather than an area.' We hope that these lessons will aid how conceptual and physical borders are managed, applied, or rethought, as disasters unfold in techno-natural ecosystems in the future.

Notes

1 In Ukraine, the name Chornobyl is used, transliterated from the Ukrainian Чорнобиль, instead of Chernobyl, which is transliterated from the Russian Чернобыль.
2 'Effective dose' is a technical term used in radiological protection. It calculates the dose for whole bodies, as opposed to individual organs (the 'equivalent dose'), which receive exposure differently.
3 This is a pseudonym used for confidentiality reasons.

References

Adey, P 2015, 'Air's affinities: Geopolitics, chemical affect and the force of the elemental', *Dialogues in Human Geography*, vol. 5, no. 1, pp. 54–75.

Alaimo, S 2010, *Bodily natures: Science, environment, and the material self*, Bloomington, IN: Indiana University Press.

Alexis-Martin, B 2020, 'Nuclear warfare and weather (im)mobilities: From mushroom clouds to fallout.' In Barry, K, Borovnik, M & Edensor, T (eds.), *Weather: Spaces, mobilities and affects*, New York: Routledge.

Amiro, B.D, Sheppard, S, Johnston, F.L, Evenden, W.G & Harris D.R 1996, 'Burning radionuclide question: What happens to iodine, cesium and chlorine in biomass fires?', *The Science of the Total Environment*, vol. 187, no. 2, pp. 93–103.

Anderson, B 2009, 'Affective atmospheres', *Emotion, Space and Society*, vol. 2, no. 2, pp. 77–81.

Anderson, B & Ash, J 2015, 'Atmospheric methods', In Vannini, P (ed.), *Nonrepresentational methodologies*, London: Routledge.

Beck, U 1992, *Risk society: Towards a new modernity*, New Delhi: Sage.

Billé, F 2017, 'Introduction: Speaking Volumes', Editor's Forum: Theorizing the Contemporary, *Society for Cultural Anthropology*, October 24, <https://culanth. org/fieldsights/introduction-speaking-volumes>.

Billé, F 2019, 'Volumetric Sovereignty: Introduction', *Society and Space: Magazine: Forums*, <https://www.societyandspace.org/forums/volumetric-sovereignty>.

Brown, K 2015, *Dispatches from dystopia: Histories of places not yet forgotten*, Chicago: University of Chicago Press.

Brown, K 2019, *Manual for survival: A Chernobyl guide to the future*, London: Allen Lane, an imprint of Penguin Books.

Brown, K 2020, 'Chernobyl is going global', *Feral Atlas*, <https://feralatlas. supdigital.org/poster/Chernobyl-is-going-global>.

Choy, T 2018, 'Tending to suspension: Abstraction and apparatuses of atmospheric attunement in Matsutake worlds', *Social Analysis*, vol. 62, no. 4, pp. 54–77.

Choy, T 2020, 'Distribution,' In Howe, C & Pandian, A (eds.), *Anthropocene unseen: A lexicon*, Punctum Books.

Collier, S.J 2013, 'Neoliberalism and natural disaster,' *Journal of Cultural Economy*, vol. 7, no. 3, pp. 273–290.

Cons, J 2017, "Seepage', from the series Speaking Volumes, Editor's Forum: Theorizing the Contemporary, *Society for Cultural Anthropology*, October 24, <https://culanth.org/fieldsights/seepage>.

Creager, A 2018, 'Human bodies as chemical sensors: A history of biomonitoring for environmental health and regulation', *Studies in History and Philosophy of Science*, vol. 70, pp. 70–81.

Davies, T 2015, 'Nuclear borders: Informally negotiating the Chernobyl Exclusion Zone,' In *Informal economies in post-socialist spaces*, Morris, J & Polese, A (eds.) pp. 225–244, New York: Palgrave Macmillan.

Elden, S 2009, *Terror and territory: The spatial extent of sovereignty*, Minneapolis: University of Minnesota Press.

Elden, S 2013, 'Secure the volume: Vertical geopolitics and the depth of power', *Political Geography*, vol. 34, pp. 35–51.

Elden, S 2017, 'Legal terrain—the political materiality of territory', *London Review of International Law*, vol. 5, no. 2, pp. 199–224.

Elden, S 2020, 'Terrain, politics, history', *Dialogues in Human Geography*, <https:// journals.sagepub.com/doi/abs/10.1177/2043820620951353>.

Engelmann, S & McCormack, D 2018, 'Sensing atmospheres', In *Routledge handbook of interdisciplinary research methods*, Lury, C, et al. (eds.), Abingdon, Oxon: Routledge, pp.187–193.

Eriksen, C 2022, 'Wildfires in the atomic age: Mitigating the risk of radioactive smoke,' *Fire*, vol. 5, no. 1:2.

Eriksen, C & Ballard, S 2020, *Alliances in the Anthropocene: Fire, plants, and people*, Singapore: Palgrave Macmillan, Pivot.

Evangeliou, N, Zibtsev, S, Myroniuk, V, Zhurba, M, Hamburger, T, Stohl, A, et al. 2016, 'Atmospheric transport of radionuclides emitted due to wildfires near the Chernobyl Nuclear Power Plant in 2015,' *Geophysical Research Abstracts*, vol. 18 (EGU2016-6034).

Evans, P 2011, 'Forest fires around Chernobyl could release radiation, scientists warn,' *The Guardian*, 26 April 2011, <https://www.theguardian.com/environment/2011/apr/26/Chernobyl-radioactive-fires-global-danger≥.

Gabrys, J 2020, 'Smart forests and data practices: From the internet of trees to planetary governance', *Big Data & Society*, vol. 7, no. 1.

Goldstein, J.E 2020, 'The volumetric political forest: Territory, satellite fire mapping, and Indonesia's burning peatland', *Antipode*, vol. 52, no. 4, pp. 1060–1082.

Grove, K 2012, 'Preempting the next disaster: Catastrophe insurance and the financialization of disaster management', *Security Dialogue*, vol. 43, no. 2, pp. 139–155.

Hao, W.M, Bondarenko, O.O, Zibtsev, S & Hutton, D 2009, 'Vegetation fires, smoke emissions, and dispersion of radionuclides in the Chernobyl Exclusion Zone,' *Developments in Environmental Science*, vol. 8, pp. 265–275.

Higginbotham, N 2019, *Midnight in Chernobyl: The untold story of the world's greatest nuclear disaster*, London: Bantam Press.

International Atomic Energy Agency (IAEA) 2014, *Radiation Protection and Safety of Radiation Sources: International Basic Safety Standards*, Vienna: International Atomic Energy Agency.

International Atomic Energy Agency (IAEA) 2020, 'Nuclear liability conventions,' <https://www.iaea.org/topics/nuclear-liability-conventions≥.

Lobo-Guerrero, L 2011, *Insuring security: Biopolitics, security and risk*, London: Routledge.

Masco, J 2006, *The nuclear borderlands: The Manhattan project in post-cold war New Mexico*, Princeton: Princeton University Press.

McCormack, D 2009, 'Aerostatic spacing: On things becoming lighter than air', *Transactions of the Institute of British Geographers*, vol. 34, pp. 25–41.

Mitchell, P 2011, Geographies/aerographies of contagion, *Environment and Planning D: Society & Space*, vol. 29, pp. 533–550.

Mycio, M 2004, *Wormwood forest: A natural history of Chernobyl*, Washington, D.C.: Joseph Henry Press.

Nading, A.M 2017, 'Local biologies, leaky things, and the chemical infrastructure of global health', *Medical Anthropology*, vol. 36, no. 2, pp. 141–156.

Nilsen, T 2020, 'Radioactive Cesium measured up north could origin from Chernobyl forest fires', *The Barents Observer*, April 21, <https://thebarentsobserver.com/en/ecology/2020/04/radioactive-cesium-measured-north-could-origin-Chernobyl-forest-fires>.

Petryna, A 2013, 'The origins of extinction', *Limn 3 Sentinel Devices*, pp. 50–53.

Petryna, A 2016, *Life exposed: Biological citizens after Chernobyl*, Princeton: Princeton University Press.

Schwartz, J.A 2006, 'International nuclear third party liability law: The response to Chernobyl,' *International Nuclear Law in the Post-Chernobyl Period*, Paris: OECD Publishing, OECD Nuclear Energy Agency and the International Atomic Energy Agency.

Shapiro, N 2015, 'Attuning to the chemosphere: Domestic formaldehyde, bodily reasoning, and the chemical sublime', *Cultural Anthropology*, vol. 30, no. 3, pp. 368–393.

Sloterdijk, P 2009, 'Airquakes', *Environment and Planning D: Society and Space*, vol. 27, no. 1, pp. 41–57.

Steinberg, T 2006, *Acts of god: The unnatural history of natural disaster in America*, Oxford: Oxford University Press

Steinberg, P & Peters, K 2015, 'Wet ontologies, fluid spaces: Giving depth to volume through oceanic thinking', *Environment and Planning D: Society and Space*, vol. 33, no. 2, pp. 247–264.

Weizman, E 2002, 'The politics of verticality', *OpenDemocracy*, <http://www.opendemocracy.net/ecology-politicsverticality/article_801.jsp>.

World Health Organisation (WHO) 2006, *Report of the UN Chernobyl forum expert group 'health'*, (eds.), Bennett, B, et al. Geneva: WHO.

World Nuclear Association 2018, 'Liability for nuclear damage', <https://world-nuclear.org/information-library/safety-and-security/safety-of-plants/liability-for-nuclear-damage.aspx≥.

Zee, J 2017, 'Downwind', from the series speaking volumes, editor's forum: Theorizing the contemporary, *Society for Cultural Anthropology*, October 24, <https://culanth.org/fieldsights/downwind>.

Zibtsev, S.V, Goldammer, J.G, Robinson, S and Borsuk, A.O 2015, 'Fires in nuclear forests: Silent threats to the environment and human security,' *Unasylva*, vol. 66, pp. 40–51.

Section V

Big data

Earth, water, air, and fire were considered fundamental constituting elements in classical times. A fifth element was also envisaged, though it gained less prominence – aether or quintessence. Aristotle described this more ethereal or spiritual element as possessing different qualities. It lacked the tactile qualities or physicality of other terrestrial elements.

In this final section, we observe a more contemporary, ethereal element that also lacks the immediacy of touch and sensation, and that we see as formative of contemporary life and lives – 'big' data. From the development of the first computers in the 1950s, the conversion of vinyl to compact discs in the 1980s, the proliferation of computers and mobile phones as essential at home and work, to the rapid transfer of all kinds of activities and processes into the online environment, digital data, and the process of digitization rapidly shapes who and how we are.

Despite its everyday significance, 'big' data can appear incomprehensible and perhaps a little magical to the untrained eye. It is based in an abstract binary number system of ones (1) and zeros (0) and dependent on service providers and devices to give it tangible form. Its capacity to slow down, run out or disappear without obvious explanation or easy resolution, renders it existentially mysterious and ofttimes, bewildering.

This diffuse invisibility co-exists in apparent contradiction to the conflation of data with information and knowledge. There is, at times, an implicit assumption that data – even 'big' data – represents concrete facts that provide a basis for reasoned calculations and decision making. Digitalization thus appears aligned with ideas of progression and progress stemming from the enlightenment.

The process of transforming all manner of things and activities into digital form took shape in the 1950s, and American insurers were early adopters of computer technology, digitising millions of life insurance policies onto UNIVAC machines (Yates 2005). These early computers used vacuum tubes, mercury columns, quartz crystals, and memory switch gates to transform information about Americans' lives, medical histories, and insurance decisions into ones and zeroes. This Digital Revolution has been equated to the Agricultural and Industrial Revolutions in its implications for culture,

DOI: 10.4324/9781003157571-18

society, and politics. It enables communications, information exchange, and trade that, in part, constitute globalization, and has been heralded as an antidote to the suppression or misuse of information by governments and corporations.

On the downside, 'big' data is associated with 'information overload' and various kinds of social isolation. It is implicated in online predation, the mass replication of unverified and unverifiable information, and wide-reaching surveillance aimed at social control. A loss of privacy through big data collection, or hacks and leaks, drives the repeated reinvention of institutional and personal online security systems.

Like earth, water, air, and fire, 'big' data can manipulate and maim, as well as liberate and sustain.

In the anticipated roll out of 'smart' digital technologies in the financial sector, insurance companies are deploying digital data in the form of 'big' data analytics (McFall & Moor 2018). Insurance technology innovations known as insurtech are heralded as a radical reconfiguration of how we understand insurance (VanderLinden et al. 2018). In mobilizing self-tracking technologies like fitness trackers, personalized data can be used to inform individualised premium prices (McFall & Moor 2018). These changes also promise to change self-perceptions as our sense of self and, perhaps, sense of place is redefined by insurer logics.

Such changes can be understood through the lens of spatial imaginaries (Watkins 2015), specifically how discourses and practices of place and space are imagined and reimagined through insurance and Insurtech, and by tracked individuals and groups. Lobo-Guerrero (2014) speculates – drawing on a fictional depiction of mass genetic testing forming the basis for insurability and thus liveability – that insurance creates a spatial imaginary of containable and manageable risk on the one hand, and uncontainable and wild uncertainty on the other. This constitutes both places and subjectivities through which embedded exclusion and fear of the uninsured and the uninsurable are used to grow enthusiasm for insurance (Booth 2020) and to maintain social order.

In this final section, Maiju Tanninen, Turo-Kimmo Lehtonen, and Minna Ruckenstein focus on the experimental rollout of personalised or 'smart' life insurance. The 'big' data underpinning this is constituted through a complexity of privacy regulations, tracking technologies, and the occasional ambivalent behaviour of those being tracked. For this data to 'work,' there is struggle and negotiation in resolving expectations with actualities, including how, for consumers, these 'big' data products come imbued with both promise and suspicion. In experimenting with the nature of their own products, insurers appear to be contributing to spatial imaginings beyond those constructed through hegemonic insurance logics. This possibility may be reflected in the shift of insurance office architecture from the grand, monumental buildings of the nineteenth century to the light and dynamic retrofitted spaces of contemporary insurtech companies. Yet, as our final author, Liz McFall, illustrates in tracking these changes, insurance companies remain an embedded part of everyday life and urban places. Thus, it appears likely – and as signposted in proceeding chapters – the spatial and temporal variegations of insurance logics may hold little transform power beyond the hegemonic.

References

Booth, K 2020, 'Insurance, and the prospects of insurability', In Wojcik, D & Knox-Hayes, J (eds.), *Routledge handbook of financial geography*, Routledge: London, pp. 400–420.

Lobo-Guerrero, L 2014, 'The capitalisation of "excess life" through life insurance', *Global Society*, vol. 28, no. 3, pp. 300–316.

McFall, L & Moor, L 2018, 'Who, or what, is insurtech personalizing?: Persons, prices and the historical classifications of risk', *Distinktion: Journal of Social Theory*, vol. 19, no. 2, pp. 193–213.

VanderLinden, S.L.B, Millie, S.M, Anderson, N & Chishti, S (eds.) 2018, *The INSURTECH book: The insurance technology handbook for investors, entrepreneurs and FinTech visionaries*, Chichester: John Wiley & Sons.

Watkins, J 2015, 'Spatial imaginaries research in geography: Synergies, tensions, and new directions', *Geography Compass*, vol. 9, no. 9, pp. 508–522.

Yates, J 2005, *Structuring the Information age: Life Insurance and Technology in the Twentieth Century*, Baltimore and London: John Hopkins University Press.

14 The uncertain element

Personal data in behavioural insurance

Maiju Tanninen, Turo-Kimmo Lehtonen, and Minna Ruckenstein

Introduction

The expectation that Big Data and Insurtech could disrupt the insurance industry has gained popularity in recent years. Insurance companies all over the world are experimenting with auto, health, and life insurance products that aim to utilise policyholders' behavioural data for various purposes, including product and price personalisation, marketing, and possibly even risk calculations (Cevolini & Esposito 2020; Jeanningros & McFall 2020; McFall 2019; Meyers 2018). These developments fall under the phenomenon of datafication, which suggests 'taking all aspects of life and turning them into data' (Mayer-Schönberger & Cukier 2013, p. 35). Today, data is everything in life that can be digitally traced: from steps, friendships, and driving habits, to breathing, purchases, and daily movements. Digital data's potential for economic value creation lies in its circulation and ability to create relations; data becomes 'lively' (Lupton 2016) in activated market relations. Thus, valuable data is potentially everywhere, but it is more uncertain in that it is 'messier' than before; it cannot be handled and confined to certain predefined uses in the same ordered way as before.

Many of the envisioned disruptive qualities of data, such as personalised pricing and individualised risk profiling, are not and will probably never be feasible because they are subject to strict regulation and contradict some of the basic mechanisms of insurance (Barry & Charpentier 2020; McFall 2019; Tanninen 2020). Yet, the potential to utilise 'messy' and 'lively' data about 'everything' (Thrift 2011) does open new prospects for insurance companies, especially regarding the insurer–insuree relationship. With behavioural data, insurers gain a new kind of access to people's lives which could allow them to develop more selective and close-knit customer relationships (Tanninen et al. 2021).

In this chapter, we look at these (potential) developments from the consumers' point of view and analyse how they experience behaviour-based life insurance products' attempts to create new kinds of data relationships. Our findings highlight the hesitation, confusion and doubt that people have towards the data practices included in the new policies. They also showcase

DOI: 10.4324/9781003157571-19

how the notice-and-consent model, utilised, for instance, in the General Data Protection Regulation (GDPR) enforced in the European Union, is an inadequate means to ensure trustworthy data practices.

Experimenting with digital data requires insurers to leave what appeared to be the ordered world of 'pure' and insulated statistical information in which they are comfortable operating. Although insurance has never been only about statistical data and actuarial calculations (Ericson & Doyle 2004; McFall 2014; O'Malley & Roberts 2014; Van Hoyweghen 2007), the ability to amass and use longitudinal data sets has been a self-evident character-istic of insurance companies to the degree that these operations have been normalised. Data has been defined by certainty in the sense that its uses and movements have been strictly regulated and predictable. However, with the new operations, insurers face novel uncertainties that involve regulatory instability and data existing 'in the wild' because it flows in the 'real world.' Before they can wholeheartedly embrace these new developments, insurers need to experiment with the promise they offer. Even if the data cannot be fitted into neat actuarial categories and statistical analyses, it is seen as a potential new tool and resource, whose value lies in correlations, probabil-ities, and predictions. Furthermore, it is hoped that digitally tracing what people do will give insurance companies visibility into their lives and offer the possibility to gently manipulate or 'nudge' (Thaler & Sunstein 2009) cus-tomers' everyday behaviour in a direction that would be more cost-efficient for insurers in the long term.

As we will demonstrate, however, all this requires that the new practices are seen as valuable and trustworthy by policyholders. If entering the messy realm of digital data is a leap of faith for insurance companies, it is equally so for their customers. Paradoxically, although insurance is intended to pro-vide security and mitigate risk, it can create new anxieties and uncertainties for the consumers (Booth & Harwood 2016). Insurance is an opaque technol-ogy to begin with, and the actual trade-offs of a given contract are difficult to estimate. Behaviour-based insurance further complicates the insurer–policyholder relationship, as activity data collected by smartwatches and smartphones and lifestyle interventions aim to gently push people towards healthier and safer habits. In other words, even if people's daily lives are already permeated by messy data practices in the realms of digital services, retail, and social media, creating new kinds of relationships with an entity like an insurance firm is far from straightforward.

To shine a light on how existing and potential policyholders see insur-ers' attempts to form relationships with them through personalised data collection, we analyse issues raised by data use through a case study of two Finnish behaviour-based life insurance policies. Our main aim is to discuss the uncertainties related to data practices. These uncertainties, we argue, are fundamental to understanding the contextual nature of datafi-cation processes. Obtaining value out of digitalisation requires that data flows can be secured; people need to trust that the operations will benefit them. For insurers, trust is a requirement for transactions, which are usually understood as an assumed aspect of the customer relationship. Our research suggests, however, that rather than being a given, trust needs to be continuously performed, situated, and embedded in everyday practices (Lobo-Guerrero 2013; Tranter & Booth 2019). In the context of behavioural insurance, it is particularly contested, as customers evaluate the degrees of trust and the overall dependability of data practices; mistrust towards the overall data ecosystem could affect the insurance policies' perceived reliability (Steedman et al. 2020).

Behaviour-based insurance is voluntary and competes with regular products in the private insurance market. Thus, consumers can choose whether to purchase a behavioural policy and submit themselves to data collection. Unlike in the world of social media, for instance, where people have entered into firm data relations, in the realm of insurance they are still considering the harms and benefits of a possible data relationship now and in the future. As Langdon Winner (1980, p. 127) argues, 'the greatest latitude of choice exists the very first time a particular instrument, system, or technique is introduced.' Below, we demonstrate the ongoing negotiations that people participate in to make sense of the data relationship with the insurance company, as it has not (yet) become intertwined with their lives; it is still easier for most people to hesitate and refuse to give up their data.

In the following sections, we first introduce our research site and methodology. Then we discuss our findings in three sections: firstly, we analyse customers' reasons for adopting and using a behaviour-based policy. Secondly, we look at how people make sense of the policies' trade-offs and what makes a 'good deal.' Finally, we discuss the doubt, hesitation, and uncertainty that new policies raise. We conclude by arguing that uncertainties related to the behavioural policies' data practices undermine their trustworthiness. Insurers, thus, need to deal with this uncertainty if they want to include 'lively' digital data in their operations.

Research methodology

Research site and focus

Our case study examines two Finnish behaviour-based life insurance policies, introduced to the market in the latter part of the 2010s by insurers we anonymise as Company X and Company Z. In Finland, citizens are provided universal health care at a very low cost and, if exposed to economic vulnerability, a decent basic income. Thus, private health and life insurance policies are often seen as a form of 'extra security' that 'supplement' the structures provided by the welfare state (Lehtonen 2014; Lehtonen & Liukko 2010). The Finnish insurance market is highly regulated as national laws, the Finnish Financial Supervisory Authority (FIN-FSA), and EU directives set limits for industry operations. Especially the GDPR restricts insurance companies' experimentations with behaviour-based personalisation in insurance (Thouvenin et al. 2019). Still, the GDPR has faced criticism for its ability to govern the current developments in the field of digital health (Marelli et al. 2020).

The new products by Company X and Company Z combine regular life insurance policies with 'smart' features, including activity tracking conducted with wristbands, smartwatches, or smartphones and eHealth services, such as online health questionnaires and coaching programmes designed by partnering companies. Data tracking is not (yet) deeply integrated into these types of insurance product or the practices of risk pooling, underwriting, and pricing. Instead, insurers frame the new services as additional benefits. For both Company X and Company Z, the policies serve as a response to recent developments, as they experiment with digital data in order to develop more engaging and personalised insurance products. At the time of the interviews, the policies of each company differed in approach. While Company X concentrated more on making available access to eHealth services and did not have an operational reward structure, such as providing premium discounts or cashbacks for active customers, Company Z's policy highlighted financial incentives: it offered its customers bonuses on their insurance coverage if they earned enough 'activity points' to fulfil certain policy requirements.

The behavioural data collected and used in the policies is generated either by tracking devices, such as activity wristbands and smart watches, or by smartphones. In both products, the data is then circulated through a health analytics company that 'purifies' the information of excessive details and glitches and selects certain variables for the insurers' use; the latter seek to collect enough data to fulfil the policies' purposes and comply with insurance regulations. By partnering with analytics companies and eHealth providers, the insurers position themselves as platforms for wellbeing services (Tanninen et al. 2021). The platform structure, however, constitutes a complex network of data relations.

Method and analysis

The empirical materials used for this article consist of 11 focus group discussions that Maiju Tanninen (MT) conducted with actual and potential customers of behaviour-based life insurance products in autumn 2017 and spring 2019. Each focus group had two to eight participants, and overall comprised 46 customers and potential customers, 24 women, and 22 men, ranging in age from their late twenties to their sixties. The discussions spanned from 45 to 90 minutes and were recorded and transcribed.

The policy customers included both people who had already held a behaviour-based policy for some time and individuals who had only recently obtained one. In addition, some informants only had a regular life insurance policy, either because they had not chosen the smart features or they had not started to use them. In fact, some of our informants had purchased a behaviour-based policy but did not remember this before being reminded of it in the focus group. Finally, MT also interviewed people who did not have life insurance policies from the companies but were seen as potential customers by the market research panels through which they were recruited. This group of informants acted as a comparison group for the insurance clients.

The data was collected in collaboration with the insurance companies as part of a larger research project. We promised to report customer insights that emerged in the focus groups to the insurance companies in order to obtain access to the field, and, especially, establish contact with policyholders, a group that is otherwise difficult, if not impossible, to reach. Because of legal restrictions, we were not allowed to recruit the customers ourselves. Instead, they were contacted by the insurance companies. This could have been a problem in terms of our results' validity if the insurers had determined the 'right' informants for us. However, as recruitment proved to be difficult, the selection of participants ended up being quite random.

The collaborative research design required MT to balance the roles of independent scholar and collaborator. For instance, she needed to emphasise in the focus groups that she did not represent the insurance company. This was generally clear to the customers, but on a few occasions, MT was still addressed as a company representative.

The preliminary analysis of the transcribed focus group discussions was conducted by MT. With the help of automated coding, MT searched for extracts which entailed the concept 'data.' After this phase of research was complete, MT carefully read the interviews and checked the selected extracts, adding or removing excerpts when needed. The selected extracts were imported into an Excel spreadsheet which MT used to conduct more precise thematic coding by hand. Through reading, comparing, and rereading, MT classified the extracts into different thematic categories that represented experiences with personal data and behaviour-based insurance. These codes included 'interest,' 'suspicion,' 'imaginary,' 'privacy,' 'reliability/ trust,' and 'user experience.' This coded data was discussed and analysed by the authors in a joint data session. The initial analysis was drafted by MT based on the data session outcomes, and the final analysis was developed jointly by all authors through rounds of writing and rewriting.

Findings

Adopting the policy

Although behaviour-based insurance policies have previously been discussed in a variety of studies (for a review, see Tanninen 2020), these have typically overlooked the policyholder's perspective. Specifically, why do people opt into these new policies and make the crucial choice of purchasing the technology? In our focus groups, people answered this question by talking about how the policy appeared to offer something interesting enough for them to acquire, though multiple reasons were provided. Tech-savvy customers were simply keen on trying out the policy, curious about its mechanisms and eager to see its future developments. Others were attracted by the self-tracking features, which they envisioned would help them understand and manage their daily routines, such as sleeping and exercise. Many informants found the policies' (potential) bonuses compelling, providing them with an opportunity to obtain extra coverage or other benefits.

Still, notwithstanding the novel features on offer, the need for insurance remained the main reason for purchasing a life insurance policy, including one with behaviour-based features. Acquiring new kinds of information on one's own life and the possibility of using self-tracking technology were seen as additional benefits, not something essential. What mattered most was the security that insurance offers. However, the 'smart' features appear to have sparked interest and affected the final decision to purchase a policy from a specific insurance provider and thus, in some cases, those features served primarily as marketing devices (see McFall 2014).

In the focus groups, a positive attitude towards and curiosity about the policies were mixed with reservations. The pronounced ambivalence should not have come as a surprise, even for Finnish insurance companies. In fact, their own market research, which was made available to us as researchers, had shown that people are generally quite apprehensive about behaviour-based life insurance products. Though people had voluntarily taken out policies, their outlook was not solely positive. Even if the informants were interested in the products and thought that they were beneficial, they remained fearful and even suspicious about the effect that the new instruments could have on policyholder privacy and on their relationship with their insurance company. Notions of smart insurance appeared to be characterised by more general 'data anxiety' (Pink, Lanzeni & Horst 2018) or 'data ambivalence' (Lomborg et al. 2020).

In the sections below, we discuss in greater detail how the customers speculated about the use of personal data in behaviour-based life insurance policies and reconciled their positive and negative feelings. The oscillation between attraction and concern is not only a characteristic of the insurer–insuree relationship but has also been documented in other kinds of data relations. In all cases, the key question has to do with boundaries: when does 'dataveillance' become too intrusive and creepy (Lupton & Michael 2017; Ruckenstein & Granroth 2020)? The informants see personal data as an asset on which they can capitalise to obtain better services and benefits. As they have chosen to purchase behaviour-based insurance voluntarily, they accept data collection. Yet, they are left with mixed feelings. People were by and large not suspicious of the precise policy that they had taken out or the company that sold it, and they generally thought that they retained their self-determination as to the degree of disclosure of their private daily routines and actions. Still, they did fear a loss of control over their personal information and struggled to make sense of the complex data relationships that these policies create.

Bargaining data

The financial incentives and rewards incorporated into behaviour-based life insurance were in principle attractive to the customers. They compared the behaviour-based instrument to car insurance products that reward accident-free policyholders with bonuses. The smart policy was seen as a similar mechanism that compensates people for staying healthy. Most of the customers that participated in the focus groups considered these reward structures to be fair. This is in part because the companies do not, at least openly, punish unhealthy or inactive policyholders. Instead, all customers retain their basic level of coverage (or premiums) and can gain bonuses (or discounts).

However, due to their experiences with the tracking devices, some customers doubted whether the self-tracked data was reliable enough for assessing activity levels and determining rewards. The inaccuracies and deficiencies of such data are widely known (Gorm & Shklovski 2019; Pink et al. 2018), and our informants also reflected on the devices' inability to measure their activities correctly; the data did not resemble their 'real selves' (Lupton 2020). Thus, even though people did not oppose the policies' rewarding structures per se, they had concerns with the trustworthiness of the behavioural data. Two of Company Y's customers, Teemu, an IT professional in his late 30s and Anne, a sales manager in her 40s expressed their concerns as follows:

TEEMU: But how they are going to measure it [health]; that is the tricky question. What data is it based on?
ANNE: Yeah, that should truly be something trustworthy. It cannot be merely the device: it's not enough.
TEEMU: Yes, it can't remain open to interpretation.

Unlike car insurance, where eligibility for bonuses is checked annually, in smart insurance the idea is that policyholders' risk scores could be assessed and determined based on real-time data (Meyers & Van Hoyweghen 2020; Zuboff 2019). However, at least in our case study, this idea appears to be unfeasible in life insurance due to both consumer objections and technological and regulatory limitations (Tanninen 2020; Tanninen et al. 2020). Many of our informants recognised that the usefulness of behavioural data stems from longer time series such as monthly averages. This was also the approach in Company Z's policy, which rewarded its customers based on their average score over a period of several months. As the final estimation was based on this longer time frame, policyholders appeared more accepting of small inaccuracies in their data.

Still, people did not deem it enough to be rewarded only after reaching the goals specified in policies. Instead, the focus groups revealed that, despite any inaccuracies, people's personal data has innate value regardless of their activity status, and a policy's terms and conditions should be attractive enough for them to give out their personal information. Clearly, people can regard their data as a form of currency with meaningful purchasing power, echoing demands made by technology developers to combat informational asymmetries. For instance, Lanier (2013) argues that as commercial agents profit from digital traces, a portion of their gains should be distributed to the data subjects as remuneration for providing their data. This view resonated with how Matti, a paramedic in his 40s, approached the matter.

> I don't think people like the idea of being monitored, or, at least, I don't like it. But if you got some support and guidance for, say, exercising – or could there be a discount for the gym, a personal trainer or dietician services [included in the policy]? I don't like the idea that in return for being stalked and monitored and being subjected to data collection and data distribution, I would get just a [premium] discount.

In the focus groups, people not only assessed existing practices but also went further. They began to imagine 'good' and 'bad' deals with insurance companies and to think about their own bargaining power. For instance, Marjo, a 45-year-old university lecturer who did not yet have a behaviour-based policy said that she 'could maybe take the smart features as a freebie if the insurance price remained the same.' Another interviewee, Eero, a chef in his 50s reflected that if he 'got a great deal with some [wellbeing] service provider,' he might allow the insurance company to gather his data. Thus, customers expected something in return for their personal data, even when they were *not* conforming to the activity or health goals set by the policies.

An especially striking finding in the interviews was that, in a world of digital services, consumers appear to value especially highly connection with, and help from real-life experts. As Matti's statement above exemplifies, people were interested in receiving guidance from medical professionals, dieticians, and personal trainers who could help them interpret their data and plan health interventions based on it. Only on some occasions did customers feel that it would be sufficient to have their data interpreted by a robot or an artificial intelligence application – a finding that must be a disappointment, considering the insurance companies' ambitions for the data economy of the near future (Grundstrom 2020). Instead of a novel, largely automated circulation of information that would enable cutting labour costs for insurance companies, our focus groups appear to imagine that the new data circuits will create more personalised services based on *human* interpretation and interaction.

The fairness of the (current) trade-offs between the data, rewards, and services was reasoned about in varying ways. Some felt that the exchange was fair, as they could get an increase in their insurance coverage or use the eHealth services attached to their policies. Others did not find the trade-off appealing enough, especially when it comes to financial rewards. This was discussed by Ossi, a customer service agent in his 30s and Hanna, a project manager in her 40s.

OSSI: A discount of five euros per year? That won't do it.
HANNA: I would just be wondering if I am selling my soul for five euros.

Obviously, small rewards neither motivate people to pursue policy goals nor compensate them for the collection of their data. Furthermore, the reference to selling one's soul for five euros vividly highlights the depth of apprehension and mistrust that people can have towards data collection. For Hanna, the actual trade-off is not clear. Will she be selling her soul to the insurance company for a relative pittance and signing up for something that might harm her?

Along with the modest financial rewards, some customers also criticised the services included in the policies. Mikko, an engineer in his 40s, said, 'the data collection is totally fine by me, but they should use it and loop it back to me so that I could get something concrete in return.' Here, the issue is not so much the mistrust placed in the data collection but the lack of a proper 'feedback loop' (Ruckenstein & Pantzar 2017) to build actionable insights with the data. As the services were not seen as advanced or tailored enough, the companies' promises of personalisation remain unfulfilled. One of the core promises of the data economy fails if the new information that is disseminated does not reach the customer in a meaningful way. Thus, instead of truly personalising prices or services, the 'smart' features only appear to help companies stand out from their competitors at the point of sale (McFall 2019; McFall & Moor 2018). Partly because of their lack-lustre experiences, a number of our informants had stopped using the policies' behaviour-based features or used them only in a desultory, unengaged way. Hence, customers were dropping out of the schemes and becoming traditional life insurance clients or, in some cases, the collection was still occurring through the mobile app without the customers' active participation or interest.

Data doubt

As the thoughts about room for bargaining above demonstrate, ideally, people want to be able to control the insurer–insuree relationship and set limits on the smart policies. The informants hoped the trade-off would be beneficial: they required something in return for their data, and some opted out of the behaviour-based services if these were not sufficiently engaging. Furthermore, they found it important to retain a sense of autonomy and feel that they *chose* the forms in which their data is tracked. Kaisa, a HR specialist in her 40s, discusses personal choice as a precondition for the decision to adopt the technology. Acknowledging the fact that she had agreed to data collection, she thinks that 'it's OK.' Yet, 'in a broader perspective,' she does not view such practices as 'a good idea,' especially if it would be 'mandatory and compulsory.' That would be too 'controlling' and too 'top-down.' More generally, our informants tended to underline the importance of smart policies and data tracking being voluntary: the data collection and 'nudging' policy features were considered acceptable if they were chosen by the policyholders.

Although people might accept the current state of a policy that they had taken out, similarly to Kaisa's case above, they feel unease regarding the smart policies' abilities and potential effects. Those possible negative effects were the subject of speculation in the focus groups, sometimes with humorous and exaggerated overtones. For instance, informants shared vivid visions of insurance companies' monitoring their behaviours, movements, and similar parameters in real time, essentially becoming unwelcome guests or even stalkers. In these exaggerated narratives, insurers would interrupt everyday situations ranging from relaxing on the sofa to having a night out by giving not only unsolicited (health) advice but also direct commands, scolding, and physically forcing the customers to return to healthy habits.

Yet, importantly, the customers were not certain which of the forms of surveillance were actually already taking place and which were only imaginary. The limits of data collection were unclear. For instance, people did not know whether the insurers received their location information and generally lacked specific knowledge of what data was being collected. This uncertainty is attested by Antti, a bank clerk in his 30s:

> Now I am not really sure which data is going there [to the insurance company]; I have just accepted that the information is transferred and which info is included. Are they [the insurance company] using just the data on the activity points? Is that enough for them, or are they receiving something else as well?

As Antti's example shows, uncertainty can exist and persist even when customers have signed an insurance contract and accepted its data policy. This doubt might be related to the policies' platform structure, as the mediation provided by the data aggregator companies and eHealth service providers complicates the data relationship. All these service providers have their own data policies for customers to accept, which makes it hard for them to keep track of who is collecting what data and all the purposes for which it is being used (Draper & Turow 2019).

The interviews made it clear that customers want to feel certain that, even if the insurance companies control the data, they would not accidentally disclose it for inappropriate uses. Despite the uncertainty related to the question of what data is being collected, the interviewees generally thought that insurance companies are trustworthy custodians of data since they have a long history of dealing with sensitive information. Still, they thought that digital data has an inherent uncertainty and is prone to security breaches (Pink, Lanzeni & Horst 2018). In a way, digital data and its movements were seen as uncontrollable, which could lead to unwelcome surprises.

For instance, the interviewees discussed the possibility that hackers could steal their data and use it for criminal purposes. They also speculated how corporate acquisitions could make their data become much more widely available than was originally intended. Moreover, people imagined how their data could come to haunt them in unexpected contexts, such as targeted advertisements, which many customers used as a reference point to make sense of the data's possible movements. Targeted advertising is something that people experience in their everyday lives: their clicks, choices, and purchases are looped back to them, sometimes creating good matches but other times resulting in annoying and even creepy encounters (Ruckenstein & Granroth 2020). Advertising is a concrete example of how personal data can be used for commercial purposes, perhaps without people being aware of it. The movements of data are just as undesirable; in the worst cases, they violate policyholders' sense of intimacy and self-determination.

Conclusion

Our study highlights the data ambivalence that is prevalent in customers' relationships with behaviour-based insurance policies and the practices those policies support. The informants were curious and interested in the products and perceived voluntary self-tracking practices not only as acceptable but also as positive. Yet, their sense of self-determination was undermined, to varying degrees, by the fact that they were not certain of what kind of data was being collected and to whom it was being made available. The analysis shows that the ambivalence extends beyond the immediate relations between people and their personal information. Uncertainties, anxieties, and apprehensions are associated with insurance, and the data economy at large, and the relationships embedded within these. Where will the data travel? Will it change the insurance terms and conditions? Will it harm me in the future?

These uncertainties undermine the policies' trustworthiness. Although people often regard self-tracked data as non-personal 'background noise' (Ajana 2020), they express concern about data movements and leakages. Our case study highlights a generalised confusion regarding what information is being collected and by whom. In practice, privacy policies are difficult to understand – even for people working in that field – and it is clearly a lot (too much) to ask people to familiarise themselves with details involved in all of their data relationships. The lack of awareness and confusion exemplifies the limitations of the notice-and-consent model used, for instance, in the General Data Protection Regulation enforced in the European Union. In light of our study, the model is inadequate in ensuring trustworthy practices, as it fails to consider people's everyday realities and hesitation when engaging with the policies (see also Marelli et al. 2020).

The customers' lack of knowledge is also related to questions of information asymmetry. The processes of datafication are built on informational asymmetries, but in the insurance context the concept usually refers to customers withholding information that is crucial for underwriting, thus increasing risk for adverse selection, that is, the disproportioned selection of high-risk individual in the pool (Baker 2003; McGleenan 1999). Social scientists have, however, pointed out that the asymmetry works the other way around, as well: insurers have much more information about a given instrument and the associated population values and averages than the customer (Van Hoyweghen 2007). Behaviour-based life insurance policies are no exception. We have demonstrated how customers struggle to make sense of the wider context of the policies and how they lack certainty on precisely what they are signing up for. Thus, the information asymmetry places policyholders in a vulnerable position, as it is very difficult for them to reliably estimate the policies' possible effects. At present, this unequal arrangement might be partly related to the policies' experimental nature; even the insurers themselves do not know what will become of the new operations and thus cannot communicate it clearly to customers (Jeanningros 2020; Meyers & Van Hoyweghen 2020; Tanninen et al. 2020).

Thus, what is at stake with uncertain data for both the insurance companies and in the data practices is how trust will be maintained or created under these new conditions. The interviewees wanted to feel secure that even if insurers (or the information technology and wellness companies that mediate the insurance practice) controlled their data, they could obtain a reasonable reward for that fact. Yet, such a transactional logic does not in and of itself guarantee trustful relations. It was hard for people to evaluate what the price of their behavioural data should be. Furthermore, customers wanted to be sure that the data would not be used for inappropriate uses such as online crime or questionable commercial practices and found it difficult to assess who to trust.

Our case speaks to the need for a careful building of trust as the insurance industry moves onto the terrain of the emerging data economy. The data relationships that insurers promote need careful planning and following through to become genuinely trustworthy. Otherwise, the industry faces the risk of raising a new kind of mistrust in people, evidence of which we can already see in the empirical material presented here. We have demonstrated how people find it difficult – if not impossible – to assess how to trust insurance, especially in the long run. If digital data is an uncertain, lively, and messy element, the insurers need to make sure that they can handle that uncertainty. Otherwise, the insurance industry as we have known it will no longer be viewed as capable of responsibly managing sensitive personal information.

References

Ajana, B 2020, 'Personal metrics: Users' experiences and perceptions of self-tracking practices and data', *Social Science Information*, vol. 59, no. 4, pp. 654–678.

Baker, T 2003, Containing the promise of insurance: Adverse selection and risk classification, In Ericson, R.V & Doyle, A (eds), *Risk and Morality*, Toronto: University of Toronto Press.

Barry, L & Charpentier, A 2020, 'Personalization as a promise: Can Big Data change the practice of insurance?', *Big Data & Society*, vol. 7, no. 1, <https://doi.org/10.1177/2053951720935143>.

Booth, K & Harwood, A 2016, 'Insurance as catastrophe: A geography of house and contents insurance in bushfire-prone places', *Geoforum*, vol. 69, pp. 44–52.

Cevolini, A & Esposito, E 2020, 'From pool to profile: Social consequences of algorithmic prediction in insurance', *Big Data and Society*, vol. 7, no. 2, <https://doi.org/10.1177/2053951720939228>.

Draper, N.A & Turow, J 2019, 'The corporate cultivation of digital resignation', *New Media & Society*, vol. 21, no. 8, pp. 1824–1839.

Ericson, R.V & Doyle, A 2004, *Uncertain business: risk, insurance, and the limits of knowledge*. Toronto: University of Toronto Press.

Gorm, N & Shklovski, I 2019, 'Episodic use: Practices of care in self-tracking', *New Media & Society*, vol. 21, no. 11–12, pp. 2505–2521.

Grundstrom, C 2020, 'Health data as enabler of digital transformation. A single holistic case study of connected insurance', PhD Thesis, University of Oulu.

Jeanningros, H 2020, 'Conduire numériquement les conduites. Économie comportementale, objets connectés et prévention dans l'assurance privée française', Thèse doctorale, Sorbonne Université.

Jeanningros, H & Mcfall, L 2020, 'The value of sharing: Branding and behaviour in a life and health insurance company', *Big Data and Society*, vol. 7, no. 2, pp. 1–15.

Lanier, J 2013, *Who owns the future?* USA: Simon & Schuster.

Lehtonen, T.K 2014, 'Picturing how life insurance matters', *Journal of Cultural Economy*, vol. 7, no. 3, pp. 308–333.

Lehtonen, T.K & Liukko, J 2010, 'Justifications for commodified security: The promotion of private life insurance in Finland 1945–90', *Acta Sociologica*, vol. 53, no. 4, pp. 371–386.

Lobo-Guerrero, L 2013, 'Uberrima Fides, Foucault and the security of uncertainty', *International Journal for the Semiotics of Law*, vol. 26, no. 1, pp. 23–37.

Lomborg, S, Langstrup, H & Andersen, T.O 2020, 'Interpretation as luxury: Heart patients living with data doubt, hope, and anxiety', *Big Data & Society*, vol. 7, no. 1, pp. 1–15.

Lupton, D 2016, *The quantified self: A sociology of self-tracking*, Cambridge, UK: Polity Press.

Lupton, D 2020, '"Not the Real Me": Social imaginaries of personal data profiling', *Cultural Sociology*, vol. 15, no. 1, pp. 3–21.

Lupton, D & Michael, M 2017, '"Depends on who's got the data': Public understandings of personal digital dataveillance', *Surveillance & Society*, vol. 15, no. 2, pp. 254–268.

McFall, L 2014, *Devising consumption: Cultural economies of insurance credit and spending*, London: Routledge.

McFall, L 2019, 'Personalizing solidarity? The role of self-tracking in health insurance pricing', *Economy and Society*, vol. 48, no. 1, pp. 52–76.

McFall, L & Moor, L 2018, 'Who, or what, is insurtech personalizing? Persons, prices and the historical classifications of risk', *Distinktion: Journal of Social Theory*, vol. 19, no. 2, pp. 192–213.

McGleenan, T 1999, Genetic testing and the insurance industry, In McGleenan, T, Wiesing, U & Ewald, F (eds), *Genetics and insurance*, Oxford: BIOS Scientific Publishers Limited.

Marelli, L, Lievevrouw, E & Van Hoyweghen, I 2020, 'Fit for purpose? The GDPR and the governance of European digital health', *Policy Studies*, vol. 41, no. 5, pp. 447–467.

Mayer-Schönberger, V & Cukier, K 2013, *Big Data: A revolution that will transform how we live, work and think*, London: John Murray.

Meyers, G 2018, 'Behaviour-based personalisation in health insurance: A sociology of a not-yet market', PhD Thesis, KU Leuven.

Meyers, G & Van Hoyweghen, I 2020 "Happy failures': Experimentation with behaviour-based personalisation in car insurance', *Big Data & Society*, vol. 7, no. 2, pp. 1–14.

O'Malley, P & Roberts, A 2014, Governmental conditions for the economization of uncertainty, *Journal of Cultural Economy*, 7, no. 3, 253–272, DOI: 10.1080/17530350.2013.860390

Pink, S et al, 2018, Broken data: Conceptualising data in an emerging world, *Big Data & Society*, vol. 5, no. 1, pp. 1–13.

Pink, S, Lanzeni, D & Horst, H 2018, 'Data anxieties: Finding trust in everyday digital mess, *Big Data & Society*, vol. 5, no. 1, pp. 1–14.

Ruckenstein, M & Granroth, J 2020, 'Algorithms, advertising and the intimacy of surveillance', *Journal of Cultural Economy*, vol. 13, no. 1, pp. 12–24.

Ruckenstein, M & Pantzar, M 2017, 'Beyond the quantified self: Thematic exploration of a dataistic paradigm', *New Media & Society*, vol. 19, no. 3, pp. 401–418.

Steedman, R, Kennedy, H & Jones, R 2020, 'Complex ecologies of trust in data practices and data-driven systems', *Information, Communication & Society*, vol. 23, no. 6, pp. 817–832.

Tanninen, M 2020, 'Contested technology: Social scientific perspectives of behaviour-based insurance', *Big Data & Society*, vol. 7, no. 2, pp. 1–14.

Tanninen, M Lehtonen, T.K & Ruckenstein, M 2021, Tracking lives, forging markets, *Journal of Cultural Economy*, vol. 14, no. 4, pp. 449–463.

Thaler, R & Sunstein, C 2009, *Nudge: Improving decisions about health, wealth and happiness*, London: Penguin Books.

Thouvenin, F, Suter, F, George, D et al. 2019, 'Big data in the insurance industry: Leeway and limits for individualising insurance contracts', *JIPITEC*, vol. 10, no. 29, pp. 209–243.

Thrift, N 2011, 'Lifeworld Inc—and what to do about it', *Environment and Planning D: Society and Space*, vol. 29, no. 1, pp. 5–26.

Tranter, B & Booth, K 2019, 'Geographies of trust: Socio-spatial variegations of trust in insurance', *Geoforum*, vol. 107, pp. 199–206.

Van Hoyweghen, I 2007, *Risks in the making: Travels in life insurance and genetics*, Amsterdam: Amsterdam University Press.

Winner, L 1980, 'Do artifacts have politics?', *Daedalus*, vol. 109, no. 1, pp. 121–136.

Zuboff, S 2019, *The age of surveillance capitalism: The fight for a human future at the new frontier of power*, New York: Public Affairs.

15 Insurance, insurtech, and the architecture of the city

Liz McFall

I realised on a business that old we had to build it outside the company. So we bought some things in Hoxton Square – I think we own about half Hoxton Square now.

(Mark Wilson, Aviva CEO 2016)

Introduction: Bubbles, bricks, and brands

Towards the end of *The Street*, Zed Nelson's 2019 film tracking the remorseless gentrification of Hoxton in the East End of London, attention turns to the building at 33–35 Hoxton Square. This, the sign outside announces, is *The Garage* and it is immediately clear that it is no longer *a* garage. *The Garage* is the UK insurer Aviva's Digital Garage, opened in 2015, in an effort to foster in-house, the kind of technological disruption traditional insurers fear coming from outside. It is no accident that Aviva called their new office The Garage, the building had functioned for many years as a commercial garage, and Aviva held on to more than the name. They also retained much of the appearance and fittings of its past functions – whitewashed brickwork, exposed services, even the garage turntable. This was all part of an effort to create the kind of edgy, industrial warehouse look favoured by tech industry start-ups, environments designed to encourage people to 'think differently.'

Aviva's Digital Garage is a neat illustration of a new direction in the kind of stories insurance companies use their buildings to tell. It is also an example of the way insurance companies, as part of their dual structure as providers of consumer products and investors in capital markets – and in real estate assets specifically – have quietly transformed urban areas. The Digital Garage is part of the Aviva 'campus' that stretches across four buildings purchased by Percussion Properties, a company set up by Aviva in 2015 and dissolved in 2018, to develop 'a campus of refurbished buildings in a vibrant and diverse London district' (Rayner 2018).

Aviva's Hoxton campus is one of the more visible examples of the role insurance enterprises play in building and rebuilding the city. The campus was designed to be visible, part of Mark Wilson's strategy, during his

DOI: 10.4324/9781003157571-20

tenure as CEO, to provide the company with a digital footprint. This use of bricks and buildings to make statements about insurance is nothing new. As I describe in this chapter, insurance companies, throughout most of their history, have relied on buildings to convey messages about the substance and weight of their intangible business. If you look, the architectural remnants of these statements can be seen in cities everywhere; vast buildings often repurposed as hotels or serviced offices. Underneath this, and much harder to see, is the way insurance has transformed the city beyond its own buildings. This is something that, excepting one or two important studies of the US context (Rubin 2009; Hanchett 2000), has barely been researched.

My aim here is to address some questions about the relationship between insurance buildings, their brands, and their broader capital, especially property, investment strategies. To do this, I review the role buildings played in the initial establishment of trust and how this began to shape the urban environment in the nineteenth and twentieth centuries. I close by looking again at the architectural precedents Aviva's garage was following. This architectural style is stock-image Silicon Valley/Alley and in two of the largest insurtech companies, Lemonade and Hippo, it is also used as a direct expression of corporate brand values. These companies have a story to tell about insurance that is quite different to the one told by legacy insurers. Their narrative is about disruption rather than stability but in using bricks to signify brands, to consumers, and more particularly investors, they stay within the bounds of a marketing strategy that has roots in the eighteenth century.

Trust in print, trust in buildings

As their corporate website chronicles, Aviva has a history stretching back to the 1690s (Aviva 2021). The website identifies their first constituent company as Hand in Hand, a mutual society that was one of the first fire insurance offices. This is roughly accurate, but it elides a few interesting details. Hand in Hand was not, in fact, the first name the company was known by; it acquired the name, as Clow (2020) explains, as an incidental by-product of the way the first fire offices conducted their business. Proposals for a system of fire insurance had circulated throughout the seventeenth century but it was only in the aftermath of the Great Fire of London that they began to take on concrete form in the shape of a scheme proposed by the Corporation of London. Before the Corporation's scheme was fully established, it was crowded out by an alternative, privately backed scheme, modelled on almost identical lines, that became the first fire office. Initially this scheme was identified only as a 'new office.' After its formal establishment in 1682, it referred to itself as the 'Insurance Office' and sometimes as the 'Fire Office' which it adopted as its official name from 1693. This name persisted until competing offices, notably The Amicable and The Friendly Society, began to be set-up prompting the Insurance Office to switch to an alternative that did not describe its function but did index one of the defining practices of the first fire insurers. This was the practice of identifying insured buildings by the presence of a firemark, a hand-sized mould of an office's seal, that was placed high on the front of the building. These firemarks were the first visual markers of insurance to appear on the streets. They identified both protected buildings and the companies they were protected by, but it was the latter function that really endured.

The Insurance Office used a phoenix on its firemark and the Amicable used the hand-in-hand symbol of mutuality. In both cases these symbols extended beyond the firemarks to become the formal identity of companies. The Insurance Office was renamed the Phoenix and the Amicable became the Hand in Hand. From the eighteenth century onwards, these symbols featured on promotional documents, on policies, letterhead, and other printed material. As Clow explains 'firemarks were objects that proceeded to create a graphic identity for the fire offices' (2020, p. 84) that lasted long after firemarks themselves fell out of use in the nineteenth century.

Graphic identities and accompanying names were well established in insurance in England by the late eighteenth century, and in Scotland a few decades later (Perman 2019). They provided a convenient means of both signifying intangible products and differentiating companies. This work was hugely significant for the fire, and later life, insurance companies[1] that proliferated in the nineteenth century (see McFall 2014; McFall & Dodsworth 2009). Rocked by, often well-founded doubts, about their propriety and their safety, many insurance companies adopted naming conventions designed to allay these concerns. In the eighteenth and nineteenth centuries, brands were built to convey solidity and respectability. They used flashes of ancient mythology – the Pelican, the Pearl and the Rock joined the Phoenix – and a succession of virtues – the Prudential, the Equitable, and many variations on the Provident and the Temperance, joined the Amicable. These names, alongside their graphic symbols, played a central role in the early development, circulation, and distribution of commercial print culture. Print, as Clow (2020) explains, was how trust in commercial enterprises was materially designed and the insurance industry devoted mountains of paper to it. It continued to matter until late in the twentieth century when the internet began to supplant it but as one, albeit central, element anchoring a broader promotional assemblage.

The proliferation of joint-stock companies in the nineteenth century further amplified the need for mechanisms that could promote trust by cultivating an appearance of reliability. Companies began to recruit, or claim to have recruited, distinguished Boards of Directors to this end. Boards were used to underwrite reputations but also to raise capital. The Duke of Buccleuch, a patron of the Scottish Widows, was repeatedly (and unsuccessfully) appealed to for investment (Perman 2019). Board directors of other Scottish insurers, including Walter Scott and the artist Sir Henry Raeburn were there partly for their glamour[2] and in Raeburn's case, also for his help in designing letterhead graphics (Perman 2019). This reputational work was circulated and repeated in an extended range of print and other materials.

By the time the flagrantly fraudulent Independent West Middlesex Fire and Life Assurance Company collapsed in 1840, print was one of several surfaces used to display corporate identity. The West Middlesex was a company that capitalised on the speculative bubble provoked at the time by the proliferation of new, joint-stock, and especially insurance companies (Taylor 2011; Alborn 2002). The West Middlesex provided the model for both Thackeray's Independent West Diddlesex Fire and Life Assurance Company in *Samuel Titmarsh* and Dickens' Anglo-Bengalee Disinterested Loan and Life Assurance Company in *Martin Chuzzlewit* (Pugsley 1991). It is Dickens' caricature though that offers the most detailed glimpse of the form insurance branding might have taken.

> The Anglo-Bengalee Disinterested Loan and Life Assurance Company started into existence one morning, not an infant institution but a grown-up company running along at a great pace, and doing business right and left: with a 'branch' in a first floor over a tailor's at the West-end of the town, and main offices in a new street in the City, comprising the upper part of a spacious house resplendent in stucco and plate-glass, with wire blinds in all the windows, and 'Anglo-Bengalee' worked into the pattern of every one of them .... Within, the offices were newly plastered, newly painted, newly papered, newly countered, newly floor-clothed, newly tabled, newly chaired, newly fitted up in every way, with goods that were substantial and expensive, and designed (like the company) to last. Business! Look at the green ledgers with red backs, like strong cricket balls beaten flat, the court guides directories, day-books, almanacks, letter-boxes, weighing machines for letters, rows of fire-buckets for dashing out a conflagration in its first spark, and saving the immense wealth in notes and bonds belonging to the company; look at the iron safes, the clock, the office seal – in its capacious self, security for anything. Solidity! Look at the massive blocks of marble in the chimney-pieces and the gorgeous parapet on the top of the house! Publicity! Why, Anglo-Bengalee Disinterested Loan and Life Assurance Company is painted on the very coal scuttles. It is repeated at every turn until the eyes are dazzled with it and the head is giddy ....
>
> (Dickens 1994, pp. 418–419)

This is a caricature of course but Dickens' description captures an environment in which branding on all sorts of surfaces had taken off. This accelerated even further when companies began to design their own offices. The Anglo-Bengalee may have been housed in a building 'resplendent in stucco and plate-glass' but it was not designed specifically for insurance. As the century progressed insurance offices, followed banks, to become among the earliest organisations to design and build their own premises. These new premises departed from the Georgian town-house style favoured by the first private banking houses in ostentatious designs, featuring commissioned religious and mythological figurative-sculptures, armorial crests and stone etched names (Rock 2008). Standard Life's premises in Edinburgh's George Street, for example, featured 'a Corinthian tetrastyle portico at centre, accommodated by projecting ground floor, with corniced doorway; fluted columns on panelled bases, frieze dated MDCCCXXV, pediment with carved tympanum representing the Wise and Foolish Virgins, by John Steell' (Figure 15.1) (Historic Environment Scotland n.d.).

Figure 15.1 a and *b* Standard Life Building George Street Edinburgh

1905

Skipper's masterpiece

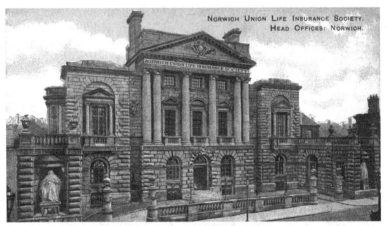

Norwich Union postcard featuring Surrey House, c. 1915

The Norwich Union Life Society's purpose-built head office, Surrey House, first opened to the public on 1 January 1905. The architect was George Skipper. The interior is richly decorated in marble, some of which was originally intended for Westminster Cathedral. Today, the building houses the Aviva Group Archive and forms an impressive main entrance for visitors. Surrey House received Grade I listed status in 1972.

Figure 15.2 Norwich Union Life Society Head Office

Once built, this architectural language was relentlessly reworked on printed material, on policies, leaflets, and stationery, and in features published in the periodical press. Figure 15.3 shows the Norwich Union Head Office at the time of its opening in 1905. Its celebration of the architect and the details of its marble interior are typical of a promotional style that endured throughout the nineteenth and into the early twentieth century (Figure 15.2).

The Norwich Union Fire Insurance Society had been formed in 1792 followed 18 years later by the Norwich Union Life Society. The two were amalgamated in 1925 and through a convoluted series of mergers and acquisitions eventually absorbed the Hand in Hand in a brand that persisted until the twenty-first century when it was renamed Aviva. Although Norwich Union was unusual in some respects, in combining general and life insurance, it was typical of the tendency to consolidation and the accumulation of vast capital sums that underpinned insurer's role as major investors in capital markets (Baker & Collins 2003; Van der Heide 2019). This process had already begun with the acquisition of property assets in the nineteenth century. For London-based insurance offices, locating in grand buildings, close to the key institutions of the Bank of England and the Inns of Court, conferred symbolic credibility and authority. It also provided practical, spatial connections to other financial institutions and facilitated participation in the transformation of the City of London into a location primarily dedicated to commerce and finance. This provided the foundation for their twentieth century role as what were often referred to as 'the institutions,' the pension and life funds that sat at 'the apex of property investment capital in the United Kingdom' (Pryke 1992, p. 239).

The 'Octopus': Insurance property investment and the city

One way of tracing how this significance to property investment came about is by considering the case of industrial life assurance, a sector that by the last quarter of the nineteenth century was the largest in both the United Kingdom and the United States[3]. Industrial assurance was targeted at the 'industrious' working classes and based on the door-to-door sale, and subsequent weekly collection, of contributions towards small life insurances used initially to meet funeral expenses. Beginning in the United Kingdom in the 1840s, with the formation of companies including Prudential Assurance, Refuge Assurance, and Pearl Assurance, by 1880 the sector was vast. In contrast to the slow trajectories of ordinary life offices, the growth of industrial offices was spectacular. A comparison of the sectors published in *The Economist* in 1892 records a rate of increase in premium income in ordinary companies in the 10 years to 1890 as 27.2%. In the same period, the rate of increase in industrial companies was 159.1%, a trend which continued with growth at over 100% in all the major industrial offices in the decade up to 1902 (The Economist 1904)[4]. By 1934, the Prudential's net Sums Assured from its Ordinary Branch[5] alone, at £25,000,000, dwarfed those of its nearest competitors in the ordinary sector, Legal and General, at £14, 539,451, and the Norwich Union, at £10,000,000 (Prudential Bulletin 1935). This exponential growth in income meant the industrial companies were amongst the largest institutional investors by the early twentieth century. When combined with the ordinary sector

> ... life insurance companies alone are estimated to have accounted for about 17 per cent of the total assets of financial institutions in the early 1920s and for some 25 per cent by the early 1960s. ... by 1938 insurance companies were one of the most important institutional investors, with assets exceeding those of the building societies by 30 per cent, and equal to two-thirds of the clearing banks' assets.
>
> (Baker & Collins 2003, p. 137)

Prudential engaged in extravagant displays of its capital as part of a promotional spectacle that became an early convention among Life Funds (Alborn
2002). Print continued to be central to this. Prudential's first colour advertising poster was produced in 1852 and displayed in railway stations across
the country (see Figure 15.3). It depicts its then head office, reinforcing the
message across media, that this was a solidly built business. As the business
grew so did its demand for premises. Its new 'chief office,' opened in 1879 at

Figure 15.3 Prudential's first advertising poster

Holborn Bars, was a loud statement to that effect. It was designed by Sir Alfred Waterhouse in a Victorian gothic revivalist style and was the largest and most expensive commission he undertook.

Prudential's headquarters dominated the Holborn landscape and it set the standard for a series of giant industrial assurance buildings across the United Kingdom, including the Royal Liver on Merseyside, Liverpool and the Refuge Assurance building in Manchester. These buildings were designed to accommodate growth and incorporated technological innovations, including telephones, pneumatic tube messaging systems, and electricity, that were becoming central to the mass data processing functions of large insurers. Still, they were never meant to function as purely practical sites of operations. These elaborately ornamented monster buildings were symbols of the capital value and economic resilience of insurance. Prudential seldom missed an opportunity to represent the building, it was photographed, drawn, and painted, appearing in policies, letterhead, posters, postcards, calendars, etc. When the company purchased £628,800 of war bonds in 1917 it had a tank parked outside Holborn Bars with the chairman on top presenting the cheque in a spectacle that was photographed, filmed, and replayed in newspapers and newsreel footage in cinemas all over the country (McFall 2014). Prudential also repeated its chief building by commissioning Waterhouse to design its branch offices. Twenty-one of these buildings were built between 1879 and 1904 – 'all featured the same trademark style and gave the company a distinctive corporate identity' (Prudential Assurance 2008) that again was amplified in print (see Figure 15.4).

This network of insurance buildings had a significant impact on the urban landscape. It can still be traced in ghost signs and inscriptions etched into the stonework of buildings that began to be sold off or leased and repurposed, often as grand hotels or serviced offices, in the late twentieth century. The effect of insurance's capital investment in property beyond their own premises, however, is much less visible. A number of studies have documented the extent of insurance institutions' investment in property in the United Kingdom (e.g. Baker & Collins 2003; Pryke 1992; Henneberry & Crosby 2015) but their focus has not been on the resulting transformation of urban environments. For that, the most relevant scholarship centres on the United States.

The Prudential Insurance Corporation of America (hereafter PIC), despite modelling itself on its UK namesake, has no formal relationship to it. Founded in 1873, its initial influence on the urban landscape, similar to the British company, came directly in the form of its own buildings and indirectly in its role as a mortgage lender. PIC stayed with the visual theme of solidity and permanence through its 'Rock of Gibraltar' trademark and hammered it home in offices that were designed to rival the scale of institutional architecture featuring 'massive promontories that both dominated and shaped their urban landscapes' (Rubin 2009, p. 3). This influence grew

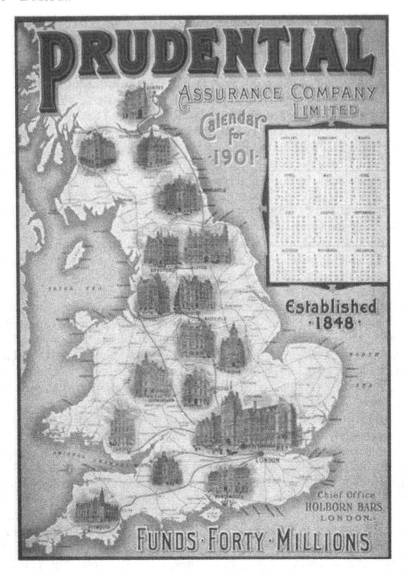

Figure 15.4 Prudential calendar 1901

in the mid-twentieth century as real estate became a financial asset that was sufficiently attractive to entice PIC, and other insurers, to not only lend but develop property themselves (Hanchett 2000). Post-World War II demographics stoked a boom in life insurance with North American company assets tripling between 1945 and 1960 and much of this landed in real estate investment.

In the three and one-half years after World War II, Prudential's real estate portfolio more than doubled, from $967 million at the start of 1946 to more than $2 billion by late 1949. In the single year 1950, Prudential made an eye-popping $1.38 billion worth of mortgage loans and property purchases. Total realty investment would surpass $7 billion by the close of the 1950s.

(Hanchett 2000, p. 315)

In the immediate post-war period, PIC had begun to invest in property development, but most of its real estate business remained concentrated in single-family mortgages (Hanchett 2000). Mortgage lending might seem a distant influence on the shape of cities but for companies that were lending at the scale of PIC, and its close competitor Metropolitan Life, lending policies had a disproportionate effect. The PIC, as Hanchett (2000) explains, did not lend in all residential districts and regarded older houses in older neighbourhoods as a greater financial risk. These restrictions;

> boded ill for American cities. When the nation's biggest mortgage maker set a policy against loans in older neighborhoods—a policy almost certainly emulated by smaller financial institutions cautiously following the giant's lead—it could became [sic] a self-fulfilling prophesy. Henceforth, even people who wished to buy older homes in the city would find it difficult to get mortgages.
>
> (Hanchett 2000, p. 317)

The influence this had on uneven development within metropolitan areas was substantial. Together with home insurance underwriting, it created a redlined bias towards the development of suburban and white neighbourhoods and against inner-city and minority communities (Squires & Velez 1987)[6]. Even so, this was far from the limits of PIC's reach.

In his 2009 dissertation, Elihu Rubin, likens twentieth century insurance to an octopus, a metaphor he borrows from Frank Norris's eponymous novel tracking the influence of railways on nineteenth century America. PIC and Metropolitan Life were two of the three largest companies in the world in 1964 and like Prudential UK, they were both founded on the weekly collection model with agents calling at 'every street, every door' (Lloyd George, quoted in McFall 2014): '[l]ike the tentacles of an octopus, Prudential insinuated itself into the social fabric of the city on a door-to-door basis' (Rubin 2009, p. 8).

The metaphor is apt. In the United Kingdom, the economic, social, and political heft of the Prudential meant it was infrastructurally entangled in the establishment of city institutions, influencing the terms of overseas trade, administering the United Kingdom's first universal healthcare payment system, partially funding two world wars, etc. In the United States, this influence also played out in the field of urban renewal with PIC's decision makers operating as planners and policymakers at both national and local levels. PIC had a vision for the cities of the future and the infrastructures that would serve them that went far beyond the design of individual developments. Focusing on the Prudential Center in Boston, Rubin describes how the company used architecture to assert its civic values and its vision for how future cities would work. This was a vision that departed from the nineteenth century city and its railway based, centralised structure towards a city that was 'multi-nodal, organised around highways and easy parking' (2009, p. 11). The Prudential Center was part of a spatial and social model of a new city that linked suburban middle-class apartments and retail centres to huge commercial offices in the inner city. PIC invested in schemes that supported this model across the United States, investing first in small suburban community centres and, from the 1950s onwards, in the development of numerous regional malls and suburban office buildings (Hanchett 2000).

The company didn't just provide the finance for all of this – its staff was directly involved throughout the organisation and administration of the development process, they helped select the tenant mix, choose architects and appointed contractors. The paradox of all of this is that PIC, as part of the broader infrastructures and institutions of insurance, were so deeply ingrained that they were everywhere and nowhere – so 'pervasively diffused through American social and economic life that they seemed to disappear into the atmosphere like so many particles of air' (Rubin 2009, p. 31). Perhaps most prophetically, in 1958, PIC helped construct the suburban Stanford Research Institute in Menlo Park, California that would, in the decades that followed, become the core of Silicon Valley, the home of the twenty-first century octopus that seems poised to disturb insurance: the tech industry.

The lightness of insurtech

In 2015, when Aviva opened its digital garage, concern about the prospect of major disruption to the business model of traditional, 'legacy' insurers from the tech industries was widespread across the sector. Management consultancy reports, press coverage, a spate of insurtech start-ups featuring Stanford trained data scientists among their founders, all signalled the threat to the industry from big data and digital technologies. By the twenty-first century insurance was dominated by giant, consolidated brands distinguished by low levels of brand differentiation, customer loyalty, and trust. Aviva was the largest British multinational general and life insurer formed from the merger of Norwich Union, Commercial Union, and General Accident and over 100 other companies. Its new brand, a palindrome based on the Latin for alive was meant to sound memorable, snappy, and global but it entered a conglomerate lexicon of equally forgettable names.

The giant returns on investment that had sustained insurance throughout the twentieth century had been on the decline since the 1990s, many of the major industrial companies had dwindled or disappeared, spectacular collapses notably Equitable's in 2000, all combined to increase the industry's vulnerability and intensify the branding challenges it faced. Huge city centre buildings and the messages they conveyed about solidity, heft, and permanence were no longer well adapted to the practice of insurance or to reassuring customers. If anything, they added to a sense of insurance as a lumbering, impersonal, and conservative industry.

This contrasts sharply with the kinds of messages insurtech start-ups want their brands to convey. Across the insurtech landscape brands are designed to be lighter, livelier, and friendlier. Brand names include Friendsurance, Brolly, Oscar, Spixii, Hippo, and Lemonade. These are brands that are not so much interested in impressing their solidity on consumers but their agility, informality, and technique. Lemonade, Inc. offers an example of how this works. Formed in New York in the same year Aviva opened its garage, Lemonade offers home, renters, and more recently, pet insurance. It was launched on the stock market with an Initial Public Offering (IPO) in July 2020[7]. In a video published on YouTube in the run-up to the IPO, one of the founders, Daniel Schreiber, introduces a series of word association customer vox pops – insurance is 'boring,' 'complicated,' 'a rip-off,' 'like a trigger word,' and 'evil.' Schreiber cuts in to explain that Lemonade's founders wanted 'to create an insurance company with an entirely different word cloud associated with it – a lot of people talk about Lemonade as being delightful, aligned, socially impactful, words like trust and love, those are the kinds of words that consumers have come to associate with Lemonade[8].' Customer delight is linked to the company's broader proposition of using machine learning, predictive data, and fast growth to transform insurance. Lemonade claims that their facility with data allows them to adopt a microscopic approach to risk.

> At twenty data points much of humanity looks alike so you price and underwrite large numbers of people as if they are a uniform, monolithic group but we found that the data we were gathering, that hundred X digital zoom we have, it showed us that groups our competitors seemed to consider to be monolithic were actually made up of predictable subgroups with over 600 percent variation in their likelihood to file a claim.[9]

Lemonade emerges in the video as a pioneer in the application of AI and BDA. This is a narrative that is targeted just as much at investors as consumers. In a reversal of the direction of investment capital, insurtechs' business model is reliant on inward not outward investment. Securing this initial funding and ongoing capital investment requires a persuasive storyline. Lemonade's storyline secured a unicorn valuation and the doubling of their share value on the first day of trading even though the company, after 5 years in business, was loss making and a very small player in its core markets (Ralph 2020). The claim that more granular, risk-based pricing will

Figure 15.5 Aviva's digital garage

be more efficient and more profitable has considerable currency among tech industry funders prepared to accept long, slow, accretive 'paths to profitability.' It is not clear that this microscopic approach to risk will lead to more accurate pricing or greater profitability (McFall, Meyers & Hoyweghen 2020) but it *is* clear that the bubbling of interest in insurtech is sustained by branding and a certain kind of material architecture (see Figure 15.5).

Insurtech companies do not look like traditional insurers. They have adopted different kinds of names, different styles of graphic identity designed to summon different 'word clouds.' Insurtech is about agility, informality, simplicity, and platform sensibility. Companies like Lemonade commission instagrammable content and employ social media to perfect conversational tones of voice. This lightness of tone is also expressed in the architecture. It is no accident that Aviva's digital venture chose Hoxton as their location or that they chose to keep many of the building's features. In doing so they followed an established path of creative tech industries moving in to 'gritty,' rapidly gentrifying urban areas formerly colonised by artists in search of large, loft or warehouse spaces in line with 'creative city' policies (Zukin 2014). This path also has an established architectural signature – bare brick, exposed services, open plan work and coffee spaces, sofas, rugs, splashes of colour. Office furnishings are modular, and walls are partitions that can easily be relocated.

The style is very much in evidence in the building interiors Lemonade and Hippo, another insurtech unicorn, feature on their corporate websites (see Figure 15.5). As Lemonade puts it, 'Our workplace looks more like a workshop than an office. We despise corporate politics, emails, and never-ending meetings. Macs are our weapons of choice, Slack is our (only) mode of communication, and we put our trust in the cloud' (Lemonade 2021). Aviva's Garage adopts the same style and has made extensive use of it to promote the company's digital credentials. At the time of its opening, Mark Wilson gave numerous interviews to the trade press (Farey-Jones 2016), the building was widely photographed, featured in architectural and design magazines and websites, and hosted dedicated tours showcasing it as an example of workplace and insurance innovation.

Everything about this vernacular suggests an airy, lightness. The Aviva Garage was fitted by SC Projects that specialises in stage deck platforms, designed to provide multi-functional, adaptable, 'edgy' workspaces.

> Employees have room to explore, develop, and test new insurance ideas and services within an environment similar to a highly flexible and personalised tech startup. [...] We supplied fifteen 1m x 1m stage decks to the Digital Garage, along with two different heights of legs to give them quick change flexibility. They have been put into use all over the building, from coffee tables to display stands in the development areas, and can be quickly altered in height to suit what's required that particular day.[10]

Insurtech, in one sense, signals a quite different landscape from traditional insurance. The emphatic architectural language of post-industrial, repurposed, retro-fitted buildings and interiors is a world away from the messages of solidity and permanence of the first purpose-built insurance offices with company names emblazoned on every available surface (c.f. Dickens 1994). In another sense, insurtech's promotional assemblage bears a strong family resemblance to its predecessors. Insurance brands, their names, graphic identities, and the specific form of their buildings are still central to the promotional assemblage and they are still designed to work together to repeat, reinforce, and amplify core messages across platforms. This work, and how it is accomplished, can have significant effects on urban environments. Aviva's investment in property around Hoxton Square is a major boost to gentrification in London's East End. This 'creative tech washing,' as Zed Nelson's film describes, changes the character of the area, raising rents and crowding out long-term residents and traditional businesses. Insurtech, like insurance, is a double facing enterprise that trades business-to-consumer (B2C) and business-to-business (B2B). The B2B market involves selling services to businesses but it also involves trade with investors and capital markets. This means engineering circuitous relationships between inward and outward investment strategies that draw on brand narratives that are expressed through the languages of print, platforms, architecture, interior design, etc. This activity shapes the character of the built environment in ways that can sometimes be impossible to miss and other times almost impossible to see.

Notes

1 Life insurance in the United Kingdom, but usually not elsewhere, was known as life *assurance*. I use the term life insurance except when referring explicitly to British companies.
2 This is a strategy that, as Geiger (2019) demonstrates, persists in some companies notably in the notorious and now collapsed digital health enterprise Theranos.
3 Industrial life assurance is distinct from mainstream 'ordinary life assurance.' Ordinary life offices did not sell industrial policies but many industrial offices sold 'ordinary branch' policies which were effectively a hybrid between the two policy types.
4 More detail on *The Economist's* figures is provided in McFall (2014).
5 All industrial companies offered 'ordinary' monthly premium policies as well as industrial policies.
6 See also the project site of Mapping Inequality: Redlining in New Deal America https://dsl.richmond.edu/panorama/redlining/#loc=5/44.34/-90.945&text=intro.
7 See https://tracxn.com/d/emerging-startups/top-internet-first-insurance-start-ups-2020 https://www.ft.com/content/58c42412-3e91-46d4-833e-1e11806c5057.
8 See https://www.youtube.com/watch?v=j7Q8SyuHWc0&t=25s.
9 Lemonade is 5 Years Old! Here's Our Strategy and Results to Date; https://www.youtube.com/watch?v=j7Q8SyuHWc0&t=25s.
10 https://scprojects.co.uk/case-study/aviva-digital-garage-london/.

References

Alborn, T.L 2002, 'The first fund managers: Life insurance bonuses in Victorian Britain', *Victorian Studies*, vol. 45, no. 1, pp. 65–92.
Aviva 2021, 'Timeline,' viewed 2 May 2021, <https://www.aviva.com/about-us/our-heritage/timeline/>.

Baker, M & Collins M 2003, 'The asset portfolio composition of British life insurance firms, 1900–1965.' *Financial History Review, vol.* 10, pp. 137–164.

Clow, M 2020, 'The design of trust, past and present: A dialogue between "design for trust" in contemporary design practice and the fire insurance industry in England 1680–1914', Unpublished PhD Thesis, School of Arts and Humanities, Royal College of Art.

Dickens, C 1994, *Martin Chuzzlewit*, Wordsworth Classics, Hertfordshire.

The Economist 1904, 'Industrial assurance', Sept 24, 1538–1539.

Farey-Jones, D 2016, 'Aviva combined marketing overhaul with business disruption', says CEO *Campaign* <https://www.campaignlive.co.uk/article/aviva-combined-marketing-overhaul-business-disruption-says-ceo/1415967>.

Geiger, S 2019, 'Silicon Valley, disruption, and the end of uncertainty', *Journal of Cultural Economy*, vol. 13, no. 2, pp. 169–184.

Hanchett, T.W 2000, 'Financing suburbia', *Journal of Urban History*, vol. 26, no. 3, pp. 312–328.

Henneberry, J.M & Crosby, N 2015, *Financialisation, the valuation of investment property and the urban built environment in the UK*, Urban Studies.

Historic Environment Scotland (n.d.), 3 George Street (Incorporating Former No 13), Standard Life, viewed 21 May 2021, <http://portal.historicenvironment.scot/designation/LB28829>.

Lemonade 2021, *Calling all makers*, <https://makers.lemonade.com/>.

McFall, L 2014, *Devising consumption: Cultural economies of insurance, credit and spending*, London: Routledge.

McFall, L & Dodsworth, F 2009, 'Fabricating the market: The promotion of life assurance in the long nineteenth-century', *Journal of Historical Sociology*, vol. 22, no. 1, pp. 30–54.

McFall, L, Meyers, G & Hoyweghen, I.V 2020, 'Editorial: The personalisation of insurance: Data, behaviour and innovation', *Big Data & Society*, vol. 7, no. 2, pp. 1–11.

Perman, R 2019, *The rise and fall of the city of money: A financial history of Edinburgh*, Edinburgh: Birlinn Ltd.

Prudential Assurance 2008, Prudential in pictures, Commemorative Booklet

Prudential Bulletin 1935, British life companies, vol. 16, no. 178, p. 2691.

Pryke, M 1992, 'Looking back on the space of a boom: (Re)developing spatial matrices in the city of London', *Environment and Planning A*, vol. 26, no. 2, pp. 235–264.

Pugsley, D 1991, 'Sham Insurance Companies Dickens, Thackeray, and the WestMiddlesex Company in Devon', 1837–1841, *Bracton Law Journal*, vol. 23, pp. 43–56.

Ralph, O 2020, 'Insurance experts put premium on start-ups reaching the big time', *Financial Times*, July 5.

Rayner 2018, 'Gentrification's Ground Zero: The rise and fall of Hoxton Square', *The Guardian*, 14 March, <https://www.theguardian.com/cities/2018/mar/14/hoxton-square-london-shoreditch-aviva-gentrification-yba-damien-hirst>.

Rock, J 2008, *Standard life buildings*, viewed 24 May 2021, <https://sites.google.com/site/joerocksresearchpages/home/standard-life-buildings>.

Rubin, E 2009, 'Insuring the city: The prudential center and the reshaping of Boston', PhD Thesis, UC, Berkeley.

Squires, G.D & Velez, W 1987, 'Insurance redlining and the transformation of an urban metropolis', *Urban Affairs Quarterly 1987*, vol. 23, no. 1, pp. 63–83.

Taylor, J 2011, 'Numbers character and trust in early Victorian Britain', In *Statistics and the public sphere: Numbers and the people in modern Britain, c. 1800–2000*, O'Hara, G & Crook, T (eds.). London: Routledge, pp. 185–204.

van der Heide, A 2019, 'Dealing with uncertainty: A historical sociology of evaluation practices in UK Life Insurance, 1971–Present', PhD Thesis, University of Edinburgh.

Zukin, S 2014/1989, *Loft living: Culture and capital in urban change*, Rutgers University Press.

16 Conclusion

Deconstructing the dualisms of elemental insurance

Chloe Lucas

In a globally changing climate, societies must find ways to live with the intensifying energies of the elements. As these energies coalesce into climatic events – such as more frequent and severe windstorms, droughts, fires, and floods – built environments are at ever greater risk of disaster. The global insurance industry has positioned itself as the key mechanism enabling society to adapt to and recover from climatic disasters. Insurance is itself a transmogrifying force, capable of turning elements into capital, and danger into profit. The authors of chapters in this book have shown that insurantial logics are pervasive and powerful: capable of reconfiguring land, water and airscapes, citizenship, sovereignty, homes, behaviours, personal relationships, and global networks. The ability of insurance to shape material and social realities can lead to transformative change in response to the climate crisis – but, as we have seen in this book, what appears adaptive in one context may be maladaptive in another.

In Chapter 2, Lauren Rickards points out that insurance responds to climate change in contradictory ways. Insurance is seen both as a pathway to changing society's relationship with our heating climate through adaptation, and at the same time as a means to defend and restore past realities. As Ewald (2019, p. 138 quoted by Rickards, Chapter 2, p. 15) puts it, insurance relates to reality by 'denying it at the same moment of its recognition.' Insurance, thus, simultaneously offers to expand into new forms and markets to enable the continuation of modern life, while threatening to collapse from overexposure to risk, leaving untold collateral damage. In this concluding chapter, we consider the capacities and limitations of insurance by examining some of the apparent dualisms of insurantial logics uncovered in this book. We synthesise the findings of these chapters to explore the relational dynamics of these dualisms: how insurance can be at once rational and emotional; technical and political; individual and collective.

Rational and emotional

As Rickards elucidates (Chapter 2), the modern insurantial imaginary enacts a culture in which recognising, measuring, and acting on risk is seen as the only rational course. In this imaginary it is not only rational but also

DOI: 10.4324/9781003157571-21

ethical for human society to aim for more complete knowledge of risk – ignorance is not regarded as an adequate defence for failing to insure. This Modern outlook, while expanding the horizons of rationality, simultaneously limits the role of emotion.

Sociologist Jack Barbalet (2019) argues that in an increasingly individuated, self-responsible culture, emotion is turned inward. Subjected to the forces of bureaucratisation and globalisation, the neo-liberalised self is re-oriented, 'not principally to consciously managing external forces but is experienced as an arena in which the individual's subjective faculties [i.e. their emotions] require self-management' (Barbalet 2019, p. 120). The opportunity to manifest and express self-transcendent emotions, which Barbalet describes as contributing to a sense of civic engagement, justice and moral duty, is consequently limited. For individuals faced with responsibility for risk in today's society, the options are therefore narrowed. Failing to purchase insurance is seen as wilfully irrational, while negative emotional responses to insurance tend to be interpreted reductively, as related to individualised emotional experience such as lack of interpersonal trust, rather than as self-transcendent commitments to different forms of just society.

Several chapters in this book explore and disturb the apparent modern dualism of emotion and rationality as it applies to insurance. Jonathon Turnbull and Christine Eriksen (Chapter 13) use the idea of affective atmospheres (Anderson 2009) to conceptualise the fear and uncertainty generated by both individuals' potential exposure to radioactive isotopes from the Chernobyl wildfires, and the inability of existing systems to measure or rationalise this risk. Other chapters show how emotion is fundamental to insurers' relationships with their customers. Liz McFall (Chapter 15) describes how insurers have responded to the affective atmospheres of the zeitgeist through the architecture of their brand. In the nineteenth century this meant creating a sense of stability and reassurance in the form of stone-faced edifices, while in the twenty-first, it means creating a visual brand for insurtech that conforms to a light-footed, innovative, hipster trend.

Trust could also be described as an affective atmosphere. Trust is both an individual and a collective phenomenon, and is both emotional and rational, in that it is a belief based on experience, whether that is direct, or indirect (Lucas et al. 2015). As Maiju Tanninen, Turo-Kimmo Lehtonen, and Minna Ruckenstein (Chapter 14) suggest, trust is easily lost unless continuously performed and embedded in everyday practices. For some insurance customers, their personal relationship with their insurer embodies the performance of trust. Bushfire survivors Jenny and Frank interviewed by Scott McKinnon, Christine Eriksen, and Eliza De Vet (Chapter 8, p. 102-103) describe their insurer as providing not only financial but also emotional support, meaning that their trust of their insurance company is distinctly interpersonal. For Northern Australian residents interviewed by Nick Osbaldiston

(Chapter 12, p. 160-161), however, 'emotional reflexivity towards insurers is built through experiences others have had, and this extends to thirdhand stories that circulate in the civil sphere.' Some of Osbaldiston's participants question the rationality of being insured, feeling that the risk of losing uninsured belongings in a cyclone helps them live 'non-materialistic lives' and is 'a good way to live.' For some the emotional excitement of 'living on the edge' is self-defining, for others a sense of fatalism together with some confidence in their own ability to get by. As McKinnon, Eriksen, and De Vet also illustrate in their chapter, the ability to rebuild and replace possessions is not commensurate with the loss of the affective, mnemonic, and emotional aspects of homes impacted by climatic disaster.

Technical and political

Actuarial science, which underlies decisions made by insurers, is the application of mathematical and statistical methods to assess risk. As Kenneth Klein points out (Chapter 9) insurance companies have enormous banks of data on which to base these statistical decisions, and so it is noteworthy when large proportions of customers, in this case victims of wildfire, find themselves unexpectedly underinsured. Policy decisions are presented by insurers as the outcome of techno-scientific calculations. Several chapters in this book examine how insurers have made political decisions behind the cover of actuarial rationalism. These authors reveal how insurantial logics that appear to be benign, technical tools for managing risks often contain hidden at their core a normative politics (or an over-riding profit motive) that can exacerbate existing social problems and inequalities.

Chloe Lucas and Travis Young (Chapter 6) describe insurantial descriptions of resilience that emphasise individual self-discipline and self-reliance, through the purchase of insurance. These normative imperatives align with the neo-liberal ideology of non-interventionist, small government. The experience of flood victims in both Hobart and Houston acted to entrench the need for self-reliance – ironically, in order to negotiate the insurance experience itself – as the application of insurance policy and practice 'acted to contradict the knowledge and testimony of the insured, and to destabilise participants' sense of control and agency in the aftermath of the disaster' (Lucas & Young Chapter 6, p. 73). The call for self-reliance and self-discipline also equates to a system that values individuals with existing wealth and privilege, while denigrating the disadvantaged. In Houston, affluent white homeowners can make this system 'work for them', while for black communities and renters, the benefit is not worth the cost.

Tanninnen, Lehtonen, and Ruckenstein (Chapter 14) describe how big data being used by insurtech is 'in the wild' – stripped of past certainties and breaking the assumptions of tried and tested technical insurance methodologies. It also offers opportunities to 'gently manipulate or 'nudge'' (Chapter 14, p. 188) customers into behaviours that suit insurance companies.

Such manipulations are inherently political, shaping people's lives to conform with individuated insurantial imaginaries. However, insurance has always been political as much as it is actuarial. Pat O'Malley (Chapter 10) describes a long-term 'fire politics' of building construction. In a recent iteration of this politics, the concept of sustainability has become technically defined in different ways for different political ends. The outcome of sustainable building, as defined by developers, is reduced building cost. This equates to less passive fire resistance, as 'embodied carbon' is stripped out. Insurers, in this instance, are on the side of long-term fire prevention, and would prefer to include more passive resistance, as they bear the financial burden of risk across the lifespan of the building.

Several chapters in this book explore how catastrophes that have historically be seen as uninsurable are rendered insurable through new forms of technocracy. As Kevin Grove (Chapter 4, p. 40) puts it, 'techniques such as risk pooling, catastrophe modelling, parametric insurance, and weather derivatives, to name but a few, have created new mechanisms for pricing and transferring risk that are based on speculative and enacted forms of knowledge, rather than actuarial and predictive knowledge.' Olli Hasu and Turo-Kimmo Lehtonen (Chapter 3, p. 25) describe how one such technique, index insurance, not only abstracts the elements it represents but transforms them into 'governable environments.' In this way, new forms of insurance technology render the previously political issue of disaster policy a technical issue, and in doing so diminish both political autonomy and democratic power. However, as Turnbull and Eriksen illustrate using the case of Chernobyl (Chapter 13), both the complex web of anthropogenic agency involved in 'natural' hazards, and the 'leaky materiality' of elements, undermine the capacity of insurance to master all forms of disaster.

Individual and collective

Throughout this chapter, I have referred to the individualising nature of modern insurance, seeing it as both a symptom and a force of neoliberal governmentalities that responsibilise individuals for collective risks. Mismatched understandings of responsibility, such as those described in Mark Kammerbauer and Christine Wamsler's example of Deggendorf (Chapter 7), mean that citizens' expectation of government protection from collective threats is undermined by individualised approaches to disaster management. Insurance seems on a trajectory to granularize and personalise policies, as described in McFall's (Chapter 15) example of an insurtech company calling itself 'Lemonade' using machine learning and predictive data to 'socially align' policies to individuals.

However, some chapters in this book describe 'flows' of risk and (social and financial) capital that undermine the idea of insurance as individual, and point towards collective forms of climate response. How well insurance can adapt to the changeable flow and interaction of risk and responsibility between individuals and groups may affect how well it can adapt to changing needs in a heating climate.

Insurance goes hand-in-hand with re-insurance, meaning that geographically specific risk pools are globalised. Zac Taylor (Chapter 11) describes, for example, how a large proportion of insurance income from high-risk residences in Florida is spread across catastrophe reinsurers in countries as far afield as Bermuda, the United Kingdom, and Germany. Taylor shows how reinsurance and insurance-linked securities displace and transmogrify climatic risks into global investment capital. But as Taylor explains, these flows are confined to regions that meet the tightly defined needs of the insurance market 'with risks that are sufficiently profitable to lure capital, actuarially well-defined enough to be priced with confidence, and where other conditions (like favourable state regulation) enable and ensure market access' (Chapter 11, p. 144).

Lack of flow (of capital) leads to accumulation in some places and deprivation in others – in essence inequality. This is the product of insurance markets such as those described by Lucas and Young (Chapter 6, p. 79) in Houston Texas, where 'insurance-driven recovery processes left renters doubly disadvantaged, often paying for rent on an uninhabitable home while searching for new or temporary residence.' Insurance increased the resilience of affluent property owners, at the expense of those with less assets. But as Rebecca Elliott (Chapter 5) notes, insurance is in essence risk sharing, in that it creates forms of 'collective mutuality' in which insurance customers pool their risk and contribute funds in order to protect one another if disaster strikes. Mutualised risk premiums, which are cross-subsidised so that people living in higher risk areas do not bear all of the cost for their risk, are seen as fair in solidarity systems, such as France, Spain, or Denmark, or as in the case of Takaful Islamic insurance (Swartz & Coetzer 2010). But in neoliberal countries with mostly privatised insurance systems, mutualisation is seen as negatively affecting climate adaptation. This is because they see granular, individualised risk-reflective pricing as sending a 'price signal' to incentivise individuals to make transformative change to reduce the underlying risk. For Elliott (Chapter 5, p. 61), the insurance industry's commitment to risk-reflective pricing is 'a core contradiction' in that addressing climatic disasters 'will require flows of resources and responsibility no matter what. Eliminating cross-subsidisation does not eliminate social interdependence.'

Insurance is seen as a principal mechanism for adaptation to a heating climate by neo-liberal governments around the world (Lucas & Booth 2020). Novel techniques for insuring catastrophic events, together with the perceived ability of re-insurers to spread risk globally are key to this belief. Big data and insurtech innovations are also promoted as transformative of the industry, enabling insurers to provide tailored policies that are also able to promote and reward adaptive behaviour. However, examples of insurance failing to provide the benefits that it promises are evidenced throughout this book.

A crucial learning from this collection of insurance research is that we need to think about insurance in less narrow ways. The centrality of social interdependence both to disaster recovery and climate change adaptation is a theme re-iterated throughout these chapters. Foregrounding the elemental – as the authors of this book have done – also signposts an understanding of human interdependence with landscapes, climate and ecology as fundamental to life in a heating world. This presents a broader way of thinking about insurance that contributes to deconstructing a suite of dualisms such as those we consider here – rational/emotional, technical/political, individual/collective. In the context of a changing climate, the intensifying energies of the elements push us to rethink insurance and with it the ways in which insurantial policy and practice contribute to patterns and trends in inequity and inequality. The elemental agency that co-constitutes insurance contributes to these dynamics in ways that indicate fruitful avenues for future inquiry. Unfolding, uncertain and rapid change is a compelling impetus.

References

Anderson, B 2009, 'Affective atmospheres', *Emotion, Space and Society*, vol. 2, no. 2, pp. 77–81.

Barbalet, J 2019, "Honey, I shrunk the emotions': Late modernity and the end of emotions', *Emotions and Society*, vol. 1, no. 2, pp. 133–146.

Ewald, F 2019, 'The values of insurance', *Grey Room 74*, vol. 74, pp. 120–145.

Lucas, C.H & Booth, K 2020, 'Privatizing climate adaptation: How insurance weakens solidaristic and collective disaster recovery', *WIRES: Climate Change*, vol. 11, no. 6, p. e676.

Lucas, C, Leith, P & Davison, A 2015, 'How climate change research undermines trust in everyday life: A review', *WIRES: Climate Change*, vol. 6, no.1, pp. 79–91.

Swartz, N & Coetzer, P 2010, 'Takaful: An Islamic insurance instrument', *Journal of Development and Agricultural Economics*, vol. 2, no. 10, pp. 333–339.

Index

Note: Page numbers in *italics* denotes figures, in **bold** tables and with "n" endnotes.

Printed in the United States
by Baker & Taylor Publisher Services